The Biodynamic Orchard Book

Also by Ehrenfried E. Pfeiffer

Pfeiffer's Introduction to Biodynamics

The Earth's Face
Landscape and its relation to the health of the soil

Weeds and What They Tell Us

The Biodynamic Orchard Book

Ehrenfried Pfeiffer & Michael Maltas

First published as *The Biodynamic Treatment of Fruit Trees, Berries and Shrubs* (by Ehrenfried E. Pfeiffer) in 1957 and *Orchard Pest Management and Spray Schedule: Biodynamic, Organic and Limited Synthetics Options* (by Michael Maltas) in 1987 by the Biodynamic Association
This edition published in 2013 by Floris Books, Edinburgh
in cooperation with the Biodynamic Association
www.biodynamics.com
Third printing 2021

www.florisbooks.co.uk
British Library CIP Data available
ISBN 978-178250-001-8
Printed and bound in Great Britain by Bell & Bain, Ltd

Contents

Publishers' Note

Ehrenfried Pfeiffer's writing on fruit trees, berries and shrubs was originally published in the 1950s, and Michael Maltas' paper was first published in 1987. Obviously since then much has changed – the organic approach is much more widely practiced and accepted now, and biodynamic farming has evolved in those years. We trust the reader will bear this in mind, and be able to see the timeless principles that these authors offer. In this spirit we have republished these two booklets together.

Floris Books and
The Biodynamic Association of North America

Fruit Trees, Berries and Shrubs

Ehrenfried E. Pfeiffer

The treatment of fruit trees, berries and shrubs should be undertaken with the aim of producing healthy plants and fruit, while avoiding the use of poisoning sprays. To make it possible to produce the desired results without any insecticides or pesticide sprays depends on many factors, not all under the control of the grower. Therefore we will discuss some of the basic principles of tree, berry and shrub biology, some of the issues the grower will encounter and some practical measures to be undertaken in the two phases of the work: the conversion period and the final goal. The grower needs to have patience and perseverance and should not expect an orchard free of pests immediately in the first or second year of the conversion phase.

The biology of trees, berries and shrubs – that is, of all wood-developing perennial plants – is entirely different from that of an annual or biennial plant. While the annual seeks its nutrients in the surface layer of the soil, the tree grows two root systems – one with feeder roots near the surface, the other sending mechanically supporting and feeder roots into the deeper layers or subsoil.

When planting trees or reclaiming older stands, both layers of the soil need attention. The preparation of an orchard begins, therefore, with the selection and pre-treatment of a suitable field.

Preparation of the soil for tree planting

If a field has never been cultivated, it has a natural structure: the humus gradually becomes less as the depth increases. Natural strata such as hardpan or clay pan or podzol layers may, however, exist. If the field was previously under cultivation there will also be all the hardpans and plow soles resulting from this. Now there is nothing a tree dislikes more than a hardpan and wet feet, that is, standing moisture in the root area, which hinders the even development of the spreading root system. Each disturbance of the root system is reflected in the growth of the tree, specifically of the crown. The results can include abnormal growth patterns, canker, bleeding and gumming, and susceptibility to pests.

The soil should be carefully prepared by deep subsoiling in order to break the hardpan and establish water and air circulation. A plant root absorbs and needs oxygen (air) for its health; it absorbs the same amount of oxygen as its own root volume every day. Poor drainage should be tackled at once. Any measure which helps establish a crumbly soil structure is an advantage. An orchard field should also be well levelled and graded; this facilitates later cultivation, mulching and irrigation.

In the first year of an orchard, subsoiling, plowing, disking and grading should be done prior to any planting. A nourishing cover crop such as vetch, rye or soya beans may be grown and disked under after well-rotted manure or compost has been spread. The more humus a soil contains, the better it is for fruit trees. Where manure or compost has been well worked into the soil, roots will develop faster. Once the trees are established, it is difficult to work in the depths of the soil; this has to be done *first*. Here we can already see some of the issues involved in reclaiming an older orchard – situations may exist with regard to hardpan and generally unfavourable structure, which cannot be overcome. No fertiliser or spray applied later on can overcome the circulatory disturbance of a tree whose root is stopped by a hardpan or standing moisture.

In shallow soils and soils with a high groundwater level, it makes little sense to select high stemmed or other trees which require a deep-growing root: choose dwarf types instead.

If you're using the biodynamic method, you will have already treated the disked-in manure and compost with preparations 502 to 507 or the biodynamic compost starter. This adds already digested organic matter and humus. Trees don't like raw manure or raw organic matter. In woods, the raw organic matter remains on the surface; only leaf mold humus is in touch with the roots. The process of humification in woods is slow but the grower has, for practical reasons, to speed it up. When the cultivation begins, the soil should be pre-treated with the biodynamic field spray or preparation 500, in order to stimulate humus formation and to activate the availability of the minerals and to encourage the fixation of nitrogen. Let us assume this has all been done correctly. (There are many books which explain this in more detail; see the Bibliography.) Then after a year of preparatory treatment, the field is ready for the planting of trees. If these things have not been done correctly, then the conditions may already have been created for many biological causes of diseases, pests and crop failure.

If you've already planted the seedling trees, you still have a chance to catch up by thorough interrow cultivation, going as close to the trees as possible without hurting the roots.

An interesting disease phenomenon was observed in Germany some years ago. Some arable fields with low productivity which, for many years, had been cultivated for crops, had been forested. Then after thirty years, the fir trees began to die off. The roots, it was found, had grown into the plow sole and other strata which had been affected by the previous cultivation.

When preparing a tree planting bed, make sure that even green manuring is well decomposed (this applies likewise for nursery beds). An important question is, should the planting hole be large and deep, or narrow? We are inclined to suggest a narrow hole just deep enough to hold the tree and its initial root development. The reason against large holes is that there is a greater amount of settling and loose soil, so that the tree tends to shake loose and 'rattle' in the hole. Also, in a large hole with looser soil, the tree may grow fast at the start. Then when the roots reach the native soil they will stop, or grow very slowly, causing circulatory disturbances in the rising sap. Cankerous growth, gum bleeding and undernourishment – that is, susceptibility of the twigs and leaves – are the consequences. We prefer a slower but steadier growth from the beginning. The tree needs a firm, well developed root system and then the necessary time for crown growth. It's better in the long run to have the first crop a year later, and have a healthy tree.

Fill the planting hole with a mixture of soil and very well rotted, earthy compost – not half-rotted or fresh compost and manure. Be sure that the soil is tightly pressed around the tree, so that it doesn't wobble in the hole. A supporting post may be used, but most trees don't need to grow 'on crutches' (the exception, of course, is if you're using the espalier method to control the trees).

Planting

Although we don't intend to contradict any particular theory, it seems to us that the best time for planting trees is early spring. Transplants which have a frozen root ball will definitely get a better start than those with dried up roots. Since roots should grow right away, we are not much in favour of transplanting in the late fall (autumn) for a root which has settled down for the winter and will not grow for some time. Then the transplant is just a stick and may rattle loose, or roots may break off during the winter.

Since many old roots will die off anyway after transplanting, it is much better to have the root system well pruned and clipped. If favourable conditions of soil, plus the urge to grow with the warming weather and the lengthening days, are provided, roots will develop at once.

Transplantation later in spring or summer is also possible if the moisture situation can be controlled. September planting may still develop roots. We are not dogmatic on the subject and we are willing to learn from other experiences. But whatever the circumstances, planting should be done under conditions which allow root growth at once. The larger the transplant – that is, the older the tree – the more important this rule is.

Should the transplants happen to arrive at an inopportune time, they should be well heeled in: nothing disadvantages the rapid start of growth more than a dried-up root.

The relationship between root stock and graft

The relationship between root stock and graft is an important one, and getting it wrong can cause untold anguish for the grower later on.

One extreme is to have a fast growing and strongly pushing root, with a slow growing variety grafted onto it. In this case the root will push a lot of sap upward and you'll get congestion in the vessels of the trunk. The fruiting area above the graft may swell up and degenerate. Knots and thickening at every joint may develop and become host to numerous infestations. In peaches and cherries, gum bleeding is frequent. Split bark can be seen.

The other extreme is a slow growing and weakly pushing root with a fast growing variety of trunk grafted onto it. Undernourishment, leaf drop and blossom drop, insufficient leaf growth, small fruit, incomplete fruit development and delayed maturing will be the result, and the fruit will not keep well. Do not believe that this series of events can be cured by fertiliser, especially nitrogen: this will only add to the trouble.

The East Malling Research Station in England has developed a 'typing' of the root stock and grafts with regard to their growth properties and the balance between root stock and tree. This grouping is called the Malling series, numbered in Roman figures I through XV. When the author talked in the United States in the 1930s about the Malling series, he found a general lack of interest, but now some people are paying attention to the problem.

The selection of the proper ***Malling type*** for a given soil and climate can help avoid many troubles later on. In fact it is paramount if one wants to have an orchard without pests and diseases. No spray or fertiliser can overcome the disadvantage of an imbalanced relationship between root and tree, that is, a disturbance in the pressure of the rising sap and its carrying vessels. When, in a human being, the circulation of the main blood vessels and of the capillaries and peripheral vessels is impaired by such diseases as arteriosclerosis (hardening) or atherosclerosis (fatty degeneration) or by other vascular diseases and spasms, serious conditions prevail. The tree also has carrying vessels which may be too narrow or too large, or the sap may rise too slowly or too fast, and serious conditions develop. One need only think a bit about the biology and physiology of sap circulation to understand this point. After all it is the sap which carries moisture and nutrients to all parts and organs of the tree, to leaves, blossoms and fruit. A leaf or blossom drop always indicates an insufficient supply of moisture and nutrients. Susceptibility to pests and disease will be the unavoidable consequence of an insufficient and imbalanced nutrient supply. We must repeat: spray guns and fertiliser bags do not make up for fundamental errors in soil structure and biological imbalance of the root and tree.

It is evident that shallow soils need a shallow-growing, spreading root, while deep soils can take a deep-growing tap root. When the tap root hits a hardpan or acid humus layer, such as a bog iron layer, it will stop growing. Even if it is able to pierce this layer, its growth will be slowed down. A disturbance of the tree results. Therefore, for ***shallow soils*** or soils which have a very high groundwater level, it is better to select only shallow-growing root stocks and dwarf trees.

For ***deep soils*** it is possible to use tall tree varieties with high trunks – long-bearing trees which grow thirty, forty, fifty or more years. These are the kinds we have inherited from our parents and grandparents. The modern economy tends toward a fast turnover and prefers dwarf varieties. These bear fruit more quickly, and are easier to harvest and to treat. But it should not be forgotten that their life cycle is short, perhaps fifteen to twenty-five years. No spray or fertiliser will make them live longer than their life cycle allows.

High stemmed trees begin to bear fruit later but live longer. If the owners of the orchard are young, or are planting an orchard for their children, they may choose a high stem and long-range program. In all other cases we advise turning to dwarf trees or medium sized ones; otherwise the orchard owners may never see the fruits of their efforts.

Another problem with regard to root stocks which is frequently overlooked is adaptability to the climate. Southern root stocks grow faster but are not hardened. Because they grow faster, many who want to see quick results try them in northern climates, but these colder climates need hardy roots. It is much better, from a biological point of view, to use northern-grown stocks and take them to a warmer climate. This also will speed growth, but at the same time there is hardiness. (Obviously this applies to the northern hemisphere.) It is also better to transplant from poorer to richer soil or from the mountain to the valley, than to do the opposite.

Under all circumstances, when buying seedling trees research the proper Malling type and choose accordingly.

Yearlings make the best transplants. They blend better and develop stronger roots in a new environment than three- to five-year-old trees. The latter may have been cut back by the nursery to look like a younger tree and the stem may look bigger than other trees of that age – but they will find it much more difficult to adjust. The apparent gain in size is not actually a gain in terms of earlier fruit.

Another difficulty in transplanting older trees – and this applies especially if full-grown trees have to be moved – is that the inner structure and biochemistry of a tree is organised according to its north-south orientation. Branches that have grown towards the southeast contain different structures and a different chemical arrangement (for instance as regards protein) than branches and leaves that grow towards the north or west. When such a tree is transplanted without regard to its original orientation, it will take several years for the tree to reorient itself. During this period of adjustment the tree is exposed to growth disturbances, is susceptible to all kinds of attacks, and in unfavourable soil and weather conditions will never recover entirely. With yearling transplants one does not have this trouble.

Finally we would like to mention the fact that north or south exposure on a hill or mountain slope also has a bearing on development. Trees on a north slope (northern hemisphere) develop a firmer, even wood structure, but grow much more slowly, and the maturing of the fruit is slower too. The ideal exposure is southeast. Reflected light – as many observations have shown – encourages significant biological activity. This is the reason why fruit grown on the north shore of a lake or river matures faster and is tastier than the same variety growing on the south side. Not everybody, of course, has a free choice in this matter.

Orchards should not be 'walled in' in a damp hole. Standing air moisture favours the development of aphids and of fungal diseases. The wind should breeze with reasonable freedom through an orchard; that said, although the stand should not be completely sheltered, some wind protection from the most violent gusts can be advantageous.

Transplanting

When transplanting, we've found it good practice to dip the root just before setting it into the new planting hole, in a liquid paste made from clay (80%) and cow manure (20%), so that the root is just covered. To this paste can be added preparation 500, which stimulates root growth. Line the planting hole with very well rotted earthy compost, mixed with the native soil and also sprayed with preparation 500. The soil should be well firmed in around the root.

If the young tree is getting a supporting pole, it is wise to tie it to the pole with a rubber band cut, for instance, from an old tire tube, in such a way that the tree is not 'hung'. Make a figure eight loop with the rubber band and place the tree in one loop and the pole into the other. The loop around the tree should be higher than the pole loop so that there is a certain 'give' to allow for settling. Nonetheless, it will in all likelihood settle a little. Ensure the tree doesn't rub against the pole.

A frequently observed error is that the soil surface in the hole, close to the trunk, is lower than the general ground level. This causes water to collect, and bark and root may rot as a consequence.

Another frequent error is that the grafting scar is buried in the soil because the tree has been planted or has settled too deep. It's useful here to discuss the difference between root bark and trunk bark. Trunk bark, which starts normally just below the grafting scar (about 1–2 in, 2–5 cm), needs air and will peel off when covered by soil, exposing the wood and causing damage to the cambium. Transplants must therefore be properly set. You can always add some earth to cover exposed root bark, but too deep a setting (in a hole) cannot be corrected by filling up with earth because then the trunk bark will be covered.

Pruning

This is not the place for a lot of detail on pruning. However, a few principal considerations should be kept in mind. The pruning of a young tree shapes it for a lifetime. If done correctly during the first two or three years, little in the way of corrective measures will be needed afterwards. The purpose of this pruning is to stimulate growth, to form a balance between vegetative growth (shoots) and fruit growth, to allow the light to enter to all parts of the branches. Branches should not be allowed to grow criss-cross and shade each other. A pyramid-shaped pattern which opens upward and outward definitely has advantages. Cross-growing twigs and branches should be removed.

New growth of one season should be cut to about seven buds from the base of the growth. It is most important that the last, outermost bud is underneath the shoot and not sideways or on top, in case the new branch grows sideward into other branches, or upward into the crown, taking away light from other branches.

Water shoots or suckers coming out from the root must be removed, otherwise they will drain away much needed sap pressure. Vertical water shoots in the crown always indicate that the tree has not been properly pruned and that there is no balance between vegetative and food growth, and no balance in sap pressure (turgor) between the root and crown. An undernourished tree also will grow water shoots in order to increase the foliage surface area. At the same time, the foliage of these shoots will take away light from the rest of the crown. Some of the best advice we ever heard came from a very experienced and successful orchardist: 'Let the tree tell you how it wants to grow and what it needs; always consider the balance, and the need for light.' Anyone who doesn't understand this should not attempt to prune. Sometimes we even find commercial tree surgeons who do not understand the growth pattern of a tree; they are tree butchers rather than surgeons.

Old trees may be rejuvenated by a severe cutting back. This is successful when done with understanding; if, the year thereafter, lots of suckers develop, you'll know that it was not properly done. Old, dead wood should be removed at once, in case the wind breaks off the branches and carries some good ones away too. Apples and pears are usually pruned towards the end of the winter, when frost can do no more serious damage, but definitely before the sap begins to rise. Peaches do well also with June pruning. This, however, is a matter of experience.

In any case, make sure the pruning-shears or knife are sharp to avoid squeezing. There should be a clean-cut face and no bark should be squeezed or torn. The cut should be as close as possible to the bud or eye, because then it will heal best. Stumps will simply die off and could lead to bark peeling and dead wood which offers a welcome nesting place for pests. The cut should be as close as possible to the vertical plane; horizontal cuts will show a 'face' exposed to the weather and lead also to bark peeling and injury. Never forget that any injury to the bark will expose the cambium and cause a serious disturbance.

Tree care

This section includes all measures, in addition to fertilisation, which make strong, vigorous growth and a healthy, resistant tree possible. Only if proper tree care is given can you then decide what the tree needs with regard to fertiliser and sprays. Tree care is frequently neglected due to the erroneous concept that fertiliser and poisonous sprays can cure all ills.

In this section we'll also discuss a suggestion on tree care made by Rudolf Steiner, the instigator of the biodynamic method.

Wherever this program of tree care has been properly applied and followed through, healthy trees have been the result and few, if any, corrective sprays have been needed. Tree care is something that just has to be done, and on the whole it's not complicated. The time and labour spent on it are well-rewarded in improved crops and a reduced need for poisoning sprays.

The tree care program below only needs to be carried out once a year, and if everything goes well it can even be only every other year. It consists of the following steps:

1. Remove all dead wood at any suitable time.
2. Prune, and cut suckers.
3. Remove all dead and loose bark, moss and lichen, by brushing with a soft wire brush. In this way you remove a lot of insect pests and their hiding places. This should not be done during the growing period when the sap is rising.
4. If the tree is otherwise healthy, washing off or spraying with a solution of preparation 500 and Equisetum tea is sufficient. It is best done in fall after harvest or in spring after pruning, and when the buds are open.
5. As a general measure for all trees which need improvement and strengthening, and as a protective measure against insect pests and disease, apply biodynamic tree paste. This is best done in fall right after the leaves have dropped, or in spring before the emergence of the buds.
6. Spray non-toxic, organic materials as needed and as often as seems necessary. It has been found, however, that the correct and thorough application of the tree paste (Step 5) will considerably reduce or even eliminate the need for sprays.

7. Biological control by way of bird protection, placing of nesting houses for birds, and introduction of predatory insects.
8. Apply biodynamic spray 501 only into the green foliage, May through July.

Tree washing with 500 and Equisetum tea (Step 4)

One portion, or unit, of preparation 500 is suspended in 2 to 4 gallons (8–15 l) of plain water or rain water. If chlorinated city water has to be used, let it stand for a few hours in a pan or bucket exposed to the daylight, if possible to sunlight. To this add a tea made from *Equisetum arvense* (horsetail). Theoretically a total solution consisting of 2% of the tea would be best. As, however, there is rarely enough equisetum available, we suggest the use of an 0.5% solution. This means that the final wash or spray solution should have a (tea) strength of 0.5%.

For each gallon of spray solution, measure out ⅔ of one ounce of the dry herb (5 g/l). Thus we would have:

for 1 gallon, ⅔ oz	for 5 liters, 25 g
for 2 gallons, 1⅓ oz	for 10 liters, 50 g
for 3 gallons, 2 oz	for 15 liters, 75 g
for 4 gallons, 2⅔	for 20 liters, 100 g

The required amount of the tea is just covered with water and brought to a boil, then allowed to simmer for 15 to 20 minutes (finely powdered or shredded *Equisetum arvense* for a shorter time, coarse material for a longer time) to make a tea concentrate. The concentrate is then mixed with the suspension of preparation 500 in water, and stirred well for about 10 minutes. Then it is sprayed into the tree so that the solution covers the entire trunk and branches. This very same spray is used as a foliage spray to reduce fungus development, especially during a wet season (damping off, or mildew, for instance). *Equisetum arvense* contains a protective factor against fungus infection. Preparation 500 stimulates the growth and renewal of the cambium – as well as doing this for the root when sprayed on the soil.

This washing of the tree is recommended in all cases where the tree has a lot of loose, peeling bark, split bark, bleeding, lesions from pruning or broken off branches, and is especially recommended if the tree is covered with mould, mildew, lichen, or moss. In the latter case it is a preparatory step to the application of the tree paste.

Tree paste application (Step 5)

In our experience this has been a most effective means – provided that the job is done right – of getting healthy trees with a smooth bark, healing lesions, and protecting the tree as much as possible against pests, especially those which hibernate underneath the bark or in crevices, like sucking insects, scale, aphids, woolly aphids, etc. The principle is that the entire tree, trunk, branches, twigs and buds, is thoroughly covered with the paste. Many of our biodynamic orchardists have covered only the trunk. This restores a healthy trunk, yet many pests hibernate and lay their eggs on the outer twigs and near the buds – for instance bud borer, aphids, scale – and in this case are not counteracted by the paste. It is especially important that not only the underside of the branches is covered, but the entire branch, including the dead corners where the branching off takes place, and that no loose bark remains to offer hiding places.

Any lesion of the timber can be painted with paste, which is a much better procedure than covering with tar, oil, asphalt, or paint, as is usually done. Holes in the trunk should be well cleaned out and then filled with paste. If eggs, larvae or scale are covered with the paste, they will be cut off from the air and they will perish. Since this paste is entirely harmless, and in no way toxic, it is an ideal means of protecting the tree and avoiding poisoning sprays. We have even sprayed it on the green foliage when this was attacked by pests and fungi (rust for instance, or mildew), so that the leaves were entirely 'painted yellow'. The rain washes it off eventually, and leaves recover with a healthy green.

The original recipe for the tree paste was: ⅓ sticky clay, ⅓ cow manure, and ⅓ fine sand. This mixture is approximate for the sticking quality varies and the proportions should be adjusted accordingly. Add as much water as is needed so that the paste can be easily applied and still stick to the tree. If needed, preparation 500, Equisetum tea, an extract of nasturtium plants against aphids, or other ingredients can be added to the solution. For many years it was biodynamic practice to apply the paste solution by hand with a whitewash brush, to the trunk and larger branches. Nowadays there are few orchardists who want to paint a tree by hand with a whitewash brush, and we admit it is a rather messy procedure. Using a pressure sprayer or spray rig is easier. For a few trees, application by hand may still be the easiest, but for a large orchard spraying is the only way to do it.

Since it has been discovered that many pests lay eggs and hibernate on the finer, outer twigs and buds, and have thus never been affected by the paste applied to the trunk, it is now our practice to cover the entire tree and even the buds with the paste in sprayable form. This is the case whether the paste is used in fall (after the leaves have fallen off), or in spring as a pre-emergence spray (before the buds open, but after the main frost period is over). In the latitude of Pennsylvania (40°N), for instance, the spring application will be at the end of February or early in March; further north it will be more toward the end of March.

For application with a whitewash brush, the clay and manure can be fairly coarse – somewhat granular. For use with spray equipment, the source materials need to be well screened in order not to block the hose, pump and nozzle. Clay is rather abrasive to pumps; apparently the piston pumps do better with it than rotary pumps. Spraying equipment such as is used for Bordeaux-lime mixtures will be the best.

The holes of the screen through which the source materials are passed before using should be of the same size as, or just a little smaller than, the spray nozzle. Use about 40 to 60 psi (3–4 bars) pressure. It is obvious that the manure and clay should be collected and prepared far ahead of time. This can be done during the off season.

Sand cannot be used in a pressure spraying system and therefore the formula needs some adjustment. Our trials were with 80% clay and 20% manure, well ground and screened so that it did not contain any fibrous material which could clog the nozzle. The degree of dilution of the paste-spray, and the timing of the spraying, have to be worked out by more trials. It is absolutely necessary that the entire tree, up into the finest twigs, is covered with paste, and no uncovered area is left. Only then will full protection be effective.

As said previously, the same spray can be applied as a foliage spray on green leaves. In practice the foliage spray is applied as a somewhat diluted solution for which any pressure sprayer will do. It is not necessary to spray this as a mist, but it should go out as fine droplets. The foliage spray can be applied several times a season, as the need arises. But under no circumstances should any spray be applied during blossoming time.

Organic sprays with no lasting residual effects (Step 6)

In heavily infected orchards, during the conversion phase, and when the danger of influx from badly infected neighbourhoods exists, one may for a time need supplementary sprays, until the general biodynamic measures and the tree paste have taken hold of the situation. There is a fundamental difference between non-toxic plant extractions such as *Equisetum arvense* (horsetail), nasturtium, herb teas or suspensions made from stinging nettle, valerian or other plant materials, (which are used as a supplementary biocatalytical or biological spray) on the one hand, and toxic organic sprays on the other. The latter, also from the plant realm, include such materials as derris, rotenone, pyrethrum, rhyania and, to a certain extent, sprays from tobacco (nicotiana).

The first group are used as strictly biological and dynamic measures – more or less 'remedies'.

The sprays of the second group can be tolerated, and even recommended, especially during the initial and conversion phases, in order to get the situation under control. The materials comprising this second group are toxic to insects and, in lesser degree, to humans if they are swallowed directly. They do not last any great length of time after application, and do not produce any lasting residues.

These organic non-lasting plant poisons are of particular interest, for insects don't seem to build up immunities to them, as they do with inorganic products. In fact, they decompose after a few days or weeks. In general, such organic plant poison sprays can be applied up to four weeks before harvest. One exception is Ryania which can be applied up to a week before harvest. Ryania, the shredded and powdered wood of a tropical shrub that grows in Trinidad, is specific against codling moth, corn borer and ear worm, and has proven to be very helpful against many other insects. Rules for application are given by the manufacturer and need not be repeated here. All these sprays should be mixed with a solvent to remove the waxy or protective layer on the skins of the pests, so that they can penetrate. They also need a glue or gummy substance, so that they stick. Soap solution, especially liquid (potassium) green soap, is recommended.

Years ago we did tests with earthworms and found that the earthworm population is not harmed or reduced by the application of these organic plant poisons, for they decompose in the soil (of course, a direct hit on a worm will kill it).

Biological control (Step 7)

For health reasons and because of the fast developing immunity of insects to toxic sprays (with the exception of the plant poisons) every attempt should be made to avoid lasting economic poisons. The first step toward achieving this goal is to establish a properly balanced humus soil. Then follows the establishment of proper biological and growth balance in the fruit tree itself. The third step is the application of non-toxic measures and tree care, as already outlined. The fourth step is biological control itself, taken by encouraging the natural enemies of the insect pests.

The most successful means is bird protection. A pair of titmice, with their young, eat 800 caterpillars a day. A cuckoo eats its own weight of insects daily. There are predatory insects like the praying mantis which lay their eggs in caterpillars and larvae, and the lady beetle whose help in destroying aphids is unsurpassable – and these are only a few of the many helpers in nature. Obviously, encouraging all these means that we need to stop spraying any poison which would interfere with their existence. So during an initial conversion phase harmless sprays only are permitted.

During the initial phase of the biological control, orchardists may have to take a year of heavy beating during the time they stop spraying poison and before the biological control is completely effective. This has to be taken in one's stride and should not discourage the grower. Once the biological control is established, things will improve. It should however be borne in mind that biological control does not take the place of the use of the tree paste described previously. The paste is needed for its beneficial influence upon the health of the tree, its cambium, its sap circulation and therefore its proper nutrition.

It has been found that 15 to 20 pairs of titmice will completely control as many acres of orchard (2½ pairs per ha). It is necessary to have nesting houses to encourage the birds to breed and settle in the orchard. These houses should have one-inch (2–3 cm) holes, so that larger birds (sparrows, for instance) cannot enter and throw out the host or steal eggs. Wren houses need even smaller holes, but the wrens will usually settle in tree holes. Needless to say, bird-catching cats do not fit into this schedule. The bird house should be cleaned during the winter and may need disinfecting (lice). Birds are always attracted when a bird bath (shallow pan) is put up for them. They need water, which should also be provided for in the wintertime, even when it freezes. If they are fed during the winter, they will prefer the place where water is also offered. And do not wait to start feeding when the frost and snow have covered the countryside, but start earlier: birds will tell each other 'this is a good place', and will stay. Otherwise a sudden, severe winter, or early snowfall, may weaken them too much to look for your place. In other words, you have to attract them before it is too late. Some suet sticks hung up here and there are very much liked. There is plenty of literature about bird protection, so we need not go into more detail.

Ladybugs (ladybirds) can be bought, but mostly they will be abundant once the biological balance has been established.

One non-toxic way to control pests is also to hang up traps for moths and other insects such as Japanese beetles. These traps are easily made from fruit jars covered with a quarter inch screen in order to shield off bees, etc. The traps are hung into the branches of the trees, and it is advisable to have them in place already at blossoming time. The bottom of a jar is filled, about 1 to 2 inches (2–5 cm) deep, with a brown sugar or molasses solution. The solution needs to be 'spiced' with an attractive aroma. For codling moths a few drops of oil of sassafras has been recommended. For other insects, the proper spice has still to be found. In these traps, aside from the moths, wasps (of the destructive kind) and Japanese beetles have been found, but only rarely a bee or a ladybug. Beehives near an orchard are, of course, essential for pollination.

Biodynamic preparation 501 (Step 8)

This preparation consists of a specially prepared, finely powdered quartz. Not much of it is needed, only a few grams per acre. These are suspended in 2–4 gallons of water per acre (20–40 l/ha) and sprayed with a fine mist-like spray into the foliage in May and June, but not while there is still any frost danger. It has to be sprayed into the green foliage and will stimulate the growth and assimilation processes of the green leaves. To stimulate assimilation means, of course, stimulation of photosynthesis, and therefore more organic nutrients for the tree, more sugar, carbohydrates and more protein.

This spray is essential if, for any reason, the tree does not develop enough foliage, or has lost its leaves from an attack of pests, fungus disease or for whatever reason. A tree which has lost its foliage because of a borer, Japanese beetle, or blight, can recover if 501 is given in June. In June (up to about June 24), the tree has the ability to open up new buds. Once the days get shorter, this is no longer possible. Preparation 501 can be sprayed, if necessary, at two-weekly intervals from the end of May to July in higher northern latitudes, in the south one could probably begin in March.

Fertilisation of an orchard

Biologically, the fruit tree stands between the forest tree and the cultivated field plant. It has the long cycle of growth in common with the forest tree. Like all forests, the orchard needs humus, and like all cultivated crops, it needs a well-aerated and fertilised soil and much care. The chief issues in orchards are the results of deficiencies, which can also come about because of over-fertilisation. These include especially deficiencies of trace minerals, and all may be absolute or relative, the latter being caused by tying down (unavailability) resulting from a onesided, unbalanced fertiliser program.

All fruit trees are sensitive to over-fertilisation with nitrogen. It is true that trees may develop more foliage because of an excess of nitrogen, and even have more fruit; but because of the imbalance the fruit may not keep as well, the blossoms may drop, diseases and pests may be fostered, and the sugar content may be reduced. In order to develop sugar, and carbohydrates in general (cellulose is a polymer), the tree needs potash. The extensive root system has more ability to get at the mineral nutrients and to make them available than does the annual plant. Nutrients therefore need not necessarily be offered in the easiest and fastest available form. This is why we prefer slower release of nutrients, as is the case in well-composted manure and compost in general. Here the soil microlife and the natural, stable humus content are important for ready availability and lasting fertility. Lasting fertility is the main factor because of the prolonged vegetation period. Quick availability, for instance in spring before emergence, would only cause more blossoms to drop, while the growing and especially the maturing fruit remains undernourished.

Humus is the balancing factor that 'holds' the nutrients as well as the moisture. The ripening fruit still needs nourishing sap at a time when most field crops are already harvested or dried up. It is frequently overlooked that a tree needs lasting fertiliser. Fresh manure is not suitable for trees for the reason that it is available too easily and its excessive nitrogen can bring bad results. In recent years a kind of nitrogen craze has developed, but it does not seem to have occurred to many so-called expert orchardists that the increase of pests and disease might be a result of unbalanced fertilisation and lack of humus supply.

Running frequently to the orchard with a spray-gun should never be considered progress. Rather, the need for so many sprays – in fact manifold times more than in the 1930s – is an alarming symptom of the disturbed balance, besides being an expense and a nuisance. It needs a reversal of thinking and habits to realise that spraying expense, which is a negative procedure, should and could be replaced, by positive, constructive action, namely soil cultivation (aeration), the building up of humus, and tree care. Only after all biological measures have been fully employed will you have healthy trees, and will the total expense of an orchard be sizably reduced. But in the meantime the orchardist should always be asking whether wasteful, negative procedures or constructive ones are being applied.

We'll now discuss some fertilisation procedures in more detail. The importance of the preparation of the seed bed has already been described. It is said that a full-grown tree in good cropping condition needs from 2 to 3 lb (1–1.5 kg) of nitrogen a year. This rule is based on the use of nitrogen fertiliser with relatively fast release of nitrogen, and on the fact that such fertilisers are about 25% efficient. To be on the safe side, we recommend compost and organic fertilisers, including composted manure and aerated sludge, containing 1 to 2 lb (0.5–1 kg) of nitrogen per full-grown cropping tree. These composts usually contain phosphates, potash, minor elements and trace minerals according to a general formula of 1–1–1 or 1–2–1 or fractions or multiples thereof. It is not necessary, when using organic fertiliser, to use a formula containing more than 2 or 3% nitrogen. It has, however, been our experience in recent years that compost in the long run will not supply all the potash that is needed.

Planting distances

Tree	*feet*	*meter*	*trees/acre*	*trees/ha*
Apple	35	11	36	88
Cherry, sour	18	5.5	135	333
Cherry, sweet	20	6	110	272
Peach	20	6	110	272
Pear	20	6	110	272
Plum	20	6	110	272

Berries	*feet between rows*	*meters between rows*	*feet apart in rows*	*meters apart in rows*
Blackberries	6	1.8	6	1.8
Black raspberries, garden	4	1.2	3	0.9
Black raspberries, field	7	2.1	4	1.2
Dewberries	4	1.2	5	1.5
Grapes	8	2.4	8	2.4
Red raspberries, garden	4	1.2	3	0.9
Red raspberries, field	7	2.1	3	0.9
Strawberries	4	1.2	1½	0.5

Trace minerals

Trace minerals should not be neglected. In compost, especially if many weeds and leaves are incorporated, all trace minerals are usually present. Tall biennial and perennial weeds, and weeds that appear chiefly during summer and fall, are richest in minerals. Nevertheless, occasionally there are deficiencies of boron, zinc, copper or magnesium.

Boron is readily available in shales, clay of marine origin and, above all, in organic matter. Its availability is reduced by liming and drought. It might increase to a toxic level by irrigation with water containing boron, for instance, in California 10 to 20 ppm in a soil is good, 4 ppm low, 30 ppm already toxic. Three to five pounds of borax per acre (3–6 kg/ha), applied as a spray or, better, mixed with compost, is sufficient. Important boron deficiency symptoms in apples are defoliation and terminal rosette formation; in shoots symptoms are rough and split bark; malformed fruit with cracks and pitting.

Manganese deficiency in apples, cherries and plums, and in black and red currants: intervenal chlorosis of leaves, beginning near margin and progressing towards midrib, all over tree, but young shoots may still show green leaves. On pears also a slightly faded green.

Magnesium deficiency on *apple*: shoot growth normal to zero, severe intervenal necrosis of leaves towards end of season, purpling, defoliation of terminal shoots, fruit fails to ripen in this case. On *cherries*: dull purple on leaves, followed by red-orange, brown necrosis in older leaves, early defoliation. On *pears:* very dark brown older leaves, intervenal necrosis near center while margin remains green (opposite to manganese deficiency). On *black currant:* central area of leaves, beginning with older ones, shows intense purple color, narrow band near margin remains green, tinted leaves curling backward, early defoliation. On *gooseberries:* slightly pale leaves, broad red margin bands, fading to cream; defoliation early. On *raspberries:* older leaves, center and margin yellow or red with green band separating. Marked chlorosis, intervenal, early defoliation. On *strawberries:* older leaves bright yellow and red, center and margin. On *grapes:* older leaves intervenal chlorosis (green grapes), or strong purple or red (blue-black varieties), narrow green margin band, brown intervenal necrosis, early defoliation.

Nitrogen and *phosphorous* deficiency need not be discussed here because they should not occur in a well-managed biodynamic orchard with sufficient organic matter supply. Occasionally there may be *iron* deficiencies. In apples and pears this may take on serious dimensions, when branches die back and fruit is highly coloured. Raspberries are quite susceptible to iron deficiency. In biodynamic soils in the United States we have not yet encountered an iron deficiency. Pale brown or black soils are suspicious in this regard.

Zinc deficiencies are mainly seen in citrus fruit. Here a foliage spray is most successful. Only minimal quantities are needed. Mottled leaves are the symptom. These show also on apples and stone fruit, as well as little leaf or rosette, accompanied by pigmentation changes, malformation of leaves, sparsity of foliage and reduced fruit production.

Copper deficiencies show up as exanthema or die back of citrus, apples, pears, plums; shoots die back, foliage shows 'burning' of leaf margin or chlorosis; rosetting and multiple buds occur, and gum bleeding takes place.

Sulfur deficiencies are rare in biodynamic soils, but may occur in sandy soils and then resemble nitrogen deficiency symptoms and sometimes show orange and red tints on the foliage of apples, gooseberries and strawberries.

Potash deficiency is important in the case of fruit trees and easily recognised in apples, gooseberries, red currants and raspberries by stunted growth, die back of shoots, bluish green leaves and especially scorching of leaves on the margin. Plenty of blossom buds, but little fruit, tasteless and of poor quality.

In plums, peaches and cherries chlorosis and forward rolling of leaves may develop in addition. Potash deficiencies, we regret having to note, are frequently seen in biodynamic and organic culture, due to an error in composting. The potash content of the compost is frequently over-estimated, and repeated applications of such composts with a low (0.5 to 0.7%) potash content have gradually caused a potash deficiency in spite of the apparent building up of humus. For growers to proceed safely, we always recommend an NPK analysis of the compost, so that they may use the proper and adequate amount, or make corrections.

Green manuring and cover crops

Green manuring and cover crops are good practice in orchards, in order to build up the effectiveness of the organic method in the soil. This brings up a frequently discussed question: open cultivation (usually with a cover crop to keep weeds down), or no cultivation at all. In fact, both approaches have their place and will fulfill their purpose if done with understanding.

In Europe there is often an overlap between a meadow or pasture and an orchard. Trees are planted into the meadow; as long as they are small, hay is mown and harvested. Under full-grown trees, cattle and sheep can pasture. Where land is scarce, for example on steep slopes that do not permit much plowing or cultivation, and on very heavy clay soils that do not yield to the plow, this practice has its benefits.

The pasturing of livestock of course becomes impossible with the use of toxic sprays. Furthermore, the increase of commercial orchards separated from other farming makes the meadow orchard obsolete. Green manuring and cover crops are therefore a substitute for the long-lasting rest period the soil would otherwise have when meadows are being developed. One should be guided by the structure and behaviour of the soil, and the needs of the trees.

There is no doubt that 'no' cultivation, as is frequently the practice in citrus orchards, can also have beneficial effects with regard to soil building. It has its place where soils are difficult to handle, organic matter is deficient, and the climate is warm and apt to destroy organic matter through heat and drought. However once the root-felt of weeds, grasses, etc., becomes so thick that no air and water can penetrate and circulate, the soil will get waterlogged or develop a separating layer between topsoil and subsoil, eventually increasing in acidity and anaerobic bacterial life. In such a case, 'no' cultivation has been overdone. Speedy turning under, breaking of the soil, and cultivation, are now necessary, followed by an alternating of 'open' years and cover or green manuring crops.

We recommend not falling into extremes but keeping a middle line. Constant open cultivation over many years will gradually break down and reduce organic matter, even to the point where it is too costly to repair the damage. This is especially the case where the soil is light, sandy (for example, in Florida), or exposed to much warmth and drought.

Clovers, alfalfa and pasture grasses can function as permanent cover crops. These have to be selected according to soil type and climate. Temporary cover crops are biennial clovers – such as mammoth clover, red clover, incarnate clover – and buckwheat on light, sandy soil for a quick turning in. An ideal combination is a mixture of red clover and mustards. Mustard sweetens out a soil but beware: it can become a terrible weed. Mustard as a cover crop should be mown in good time, so that it cannot go to seed. The hay can be used for mulching or composting; the compost can then be returned when the cover crop is plowed or disked under for a fallow period.

It is always good to top dress the cover or green manuring crop with compost immediately prior to turning under. If, at the same time, the biodynamic field spray is applied, the green manuring crop will be broken down fast without tying up nitrogen. Several cultivations can follow and then a few months or a year of plain fallowing.

If irrigation ditches have to be established and water runs maintained, the practice of green manuring needs to be well timed in order not to interfere with irrigation. Sprinkler systems have the advantage that they do not interfere with cultivation. Portable aluminum pipes have helped solve this problem. Care should be taken, though, that the sprinkler system does not spray and hit the foliage on a hot and dry day. The water should be sprinkled *underneath* the tree. Drenching with water should be avoided. Sprinkling with large drops of water under force may harden and encrust an open soil surface – in which case surface cultivation should follow to break the crust (or a finer spray should be applied). One can very well irrigate green manuring crops, but the cultivation, mowing, and irrigation of cover crops needs a level or well-leveled graded land with as little unevenness and as few ridges as possible. This grading should be done before the orchard is planted: later on it will be difficult.

Cover crops are much needed during the interim phase of young growth when the trees do not as yet shelter the soil. Interspersing crops with vegetables is possible at this time too.

Now a word about mulching. We must draw a distinction between a merely protective mulch, and nourishing mulch. The function of the protective mulch is just to cover, to hinder moisture losses and weed growth. This can be done with anything which covers, even paper; chopped straw or hay are best but you can also use salt hay, leaves, bagasse (stays dry), sawdust or wood chips – the two latter produce an acid soil. After its function is fulfilled, the mulch can be disked under or else it should be removed. New mulch should not remain over the winter because mice and rats can build tunnels underneath it and attack roots and trunks. Furthermore, the frost action is so beneficial to soil structure that it should not be hindered by letting the mulch lie over winter in a northern climate.

Mulch used in order to hinder evaporation during the hot and dry season is quite beneficial and important, but mulch in wintertime is a double-edged affair: if it hinders moisture from reaching the ground and penetrating, it has missed its purpose.

No mulch should ever come near the trunk and grafting scar of the tree. Air should have access at all times to these places, otherwise serious damage to the air bark will occur and pests may hibernate there. As a general rule, mulch should be kept one to two feet (30–60 cm) away from the tree. Much damage has been done by not observing this rule.

The second type of mulch, as mentioned above, is nourishing mulch. This is made with half-rotted compost. It should not be applied as thickly as the straight cover mulch. The compost should have lost at least some of its 'rawness' (decay of fresh material), especially when green plant parts of vegetables are used. Leaves need not be well rotted for mulching. This nourishing mulch may last only a few months, and should be disked under after it has fulfilled its function. This is best done together with an application of the biodynamic field spray. Such a mulching procedure will not completely fulfil the nutritional needs of the trees or berries, and requires a compost supplement. The compost is applied on top of the mulch immediately prior to the plowing or disking under.

We cannot advocate so-called sheet composting of raw materials, for it is apt to spread disease and vermin. Neither orchards nor vegetable gardens are proper sites for spread-out garbage dumps. Excessive mulching, packed down, can restrict air circulation and choke out feeder roots underneath the surface. We remember one case specifically where the grower had applied a heavy mulch of sludge. Digging a soil profile, we found the feeder roots were not reaching the surface soil but were some 8 to 10 inches (20–25 cm) below the top. There was lots of moisture and humus just below the mulch, but the soil strata had separated and no nutrients of the topsoil would reach the feeder roots.

Whether your mulching and cultivation practice is right or wrong is something you can easily detect when digging for feeder roots. These should come up to the surface. If they don't, something is wrong. The more feeder roots nearer the surface, the better.

There is one mulch that is especially beneficial: alfalfa hay. This hay can be used as is, preferably shredded or chopped in one to two inch (3–5 cm) lengths and slightly rotted. Old trees can be regenerated and many deficiencies corrected with it. The expense of such hay is money well invested. We met an eighty-year-old citrus grower who told us that his father had used the alfalfa mulch as a remedy, and he did so too when trees began to show poor foliage and small fruit. This mulch need not be applied every year but only once in a while, perhaps every five or ten years. It will rejuvenate the trees. This is a dual purpose mulch, both protective and nourishing. Once the hay is completely rotted, the remainder can be disked under.

Pollination of blossoms

Bees, bumblebees and other beneficial insects belong in an orchard. Less well known is the necessity for 'rooster' trees, which supply the pollen for cross-pollination. Good nurseries have lists available of 'rooster' trees for the varieties being planted. Unfortunately no suggestion regarding varieties can be made here, for this question depends too much on soil, climate, local conditions and, last but not least, individual taste and market requirements.

Mixed culture instead of monoculture

It is evident that a pest will spread much faster and more easily if it is provided with good hunting ground. If you have ten, thirty or a hundred acres of one variety, pests will spread out much faster. Insects are highly specialised and certain ones will attack one type of fruit more than another and one variety more than another. The more mixed the stand is, the better for its biodiversity and protection against pests. A completely mixed stand of different varieties and different species is biologically certainly the healthiest. Alas, economic considerations, labour, and practices of spraying and cultivation, interfere with the ideal.

We can see the biological advantage of a mixed stand in a forest, where hard woods and soft woods together grow better and produce better humus than a monoculture. The roots penetrate into different strata of soil and support rather than compete with one another. Also there are various beneficial and antagonistic interactions between root systems, even between the trees and annual plants growing around them. Nasturtium, growing between fruit trees, transmits a 'flavour' to the tree that is entirely disagreeable to aphids. (Even a washing of a tree with nasturtium juice and especially of the internodes of the twigs, where the aphids nests, helps to protect against them.)

Much too little is known as yet about beneficial or hindering interactions, otherwise we would make more definite suggestions. We would like to encourage readers to make observations in this field and report them to their local biodynamic associations.

For practical reasons a compromise between monoculture and mixed culture can be made: plant double, triple or quadruple rows of one variety or species, and alternate with double (etc.) rows of other varieties and species. For instance, set 4 rows of apples subdivided into 2 rows each of two different varieties, then 2 to 4 rows of pears, then 4 rows of peaches. The spray rig going through the row, or the harvest truck, or any cultivation process, can be used simultaneously for the double row of a kind.

The crown and the root system

It's helpful to realise that the root system is a replica of the tree crown. As the tree branches out, so the roots fan out. Dying back from the tips of the shoots and twigs indicates damage of the roots. Recent research has shown that on naturally growing trees, the individual root branch supplies the tree branch directly above it, the sap carrying vessels which lead straight up to that branch. On grafted trees this anatomical correspondence is less evident. The feeder roots are, or should be, right underneath the crown drip (the outer edge of the crown) and from there on outward. No feeder roots (of a full-grown tree) are near the trunk. Any application of compost, mulch, etc., and any cultivation or aeration, even irrigation, should therefore be applied in a space from about 2 feet (60 cm) inside to 2 feet outside of the crown drip. All these measures should stay at least 2 feet away from the trunk in older trees, and 1 foot (30 cm) in younger trees.

The cultivation of the interspace between rows, or the cultivation ring around solitary trees is, therefore, important. Whether the orchard is an orchard-meadow combination, using cover crops, or no cultivation, you could choose to alternate the approach each year, selecting alternating rows: one year nearer to the tree, one year further to the outside of the crown drip, going through the orchard in a checkerboard fashion.

Orchardists should be fully aware of the root pattern of their trees in order to apply the right practice at the right time. Apple trees, for instance, grow a deep-growing tap root for the first nine years. Then the horizontal branching begins to fan out. Mulch and compost should by that time be placed outside the crown drip, to encourage the spreading feeder roots. The pear tree has a closer system of roots, and mulch and compost spread underneath and inside the crown drip will be more effective. If a tree is strong on one side and under-developed on the other, it means also that the root is under-developed on the weak side.

Now a word about special care of solitary trees (of any kind) that may stand on the lawn, near a sidewalk, or otherwise in a position which does not allow for mulching or cultivation. This may perhaps be a tree with special associations, which one wants to rescue. In such cases we recommend making holes with a crowbar, just inside and outside the crown drip. They should be about 18 inches (45 cm) deep, and widened to about 2 inches (5 cm) in diameter by poking and twisting with the bar. Make these holes all around the tree, at distances of 2 feet (60 cm) from one another, and then fill them with well-rotted compost.

Plant diseases

A few basic precautions can significantly reduce the incidence of plant diseases. Firstly, stagnant air moisture is to be avoided: the wind should blow freely through the orchard. A closed-in stand will always cause difficulties. Secondly, the trees should be well trimmed and get all the light they can through all parts of the crown. Thirdly, too close a stand or one too near a wall or house will encourage fungus infection, mildew, blight, rust, etc. Trees and shrubs are frequently too close to a house; probably the planter did not realise the size of the full-grown tree or shrub when planting. Espalier should stand at least a few feet away from the wall and not be trained along the wall, but on a free-standing support. Roses and vines will be the more mildewed the nearer to a wall they grow, for they cannot take the reflected heat and the moisture which builds up there. (The latter is not good for the wall either.) Grapevines especially are sensitive to standing and excessive air moisture. A location in a damp hollow in the reflecting area of a lake has a lot to do with such a condition. Again, rambler roses on the southeast corner of a house will be mildewed, while on any other corner they may do well. As we have said before, the spray-gun does not solve the problem of improper position.

Fourthly, no dead, fallen fruit should ever be allowed to remain underneath the tree, for it will always spread infection. Such fruit, as well as disease-dropped leaves, should be composted, and the compost should be well decomposed and earthy before it is reapplied to the orchard. Monilia-infested fruit frequently remains on the tree, shrivels and hangs there still the next year. All monilia-infested fruit should, under all circumstances, be removed at once and be burnt.

Fortunately, industrial sprays against diseases have not been developed to the extent of the insecticides and so growers are more likely to seek biological solutions, in the absence of a suitable poison.

Some organic phosphorous and fluoride compounds are offered for use against plant diseases. They have the disadvantage that they are poisonous to enzymatic processes in humans, animals and in the plant itself. As stated before, biological control is preferable to chemical control. Orchardists should be much more aware of the relationship between the host of fungus diseases and the infected tree or shrub. The most famous example is in another field of agriculture: the stem rust of wheat, oats, rye and grasses, for which the Berberis (barberry) shrub is the alternate host. No such shrubs, therefore, should be allowed in the neighborhood of grain fields. In orchards we have cedar-apple rust, with a red cedar as alternate host, and blister rust on five-needle pines, for which currants and gooseberries are the alternate host.

Fifthly, poor and unprotected pruning cuts, and any bark lesions, are particularly susceptible to fungus invasion and rot, and should be immediately covered with tree paste (see page 22). Canker on apples and pears is caused by a fungus, *Septobasidium pedicellatum* – use the tree paste. White rot appears on fruit, nut and shade trees, hence the latter, if in the neighborhood of fruit and nuts should also receive biological treatment. The mycelium and perithecia of the apple scab fungus grow and propagate in dead leaves, so sheet composting and mulching with dead leaves are great ways to spread the scab fungus. All dead leaves should be collected, as well as dead, dropped fruit, and be very well composted to an earthy compost structure. Leaf spot on cherry and plum, blight on stone fruits, grey mold or botrytis blight on buds, fruits and roots, blue mould on apple, pear or citrus, are caused by fungi imperfecti (Deuteromycota) and can be combated with tree paste and Equisetum spray.

Since we do not object to sprays of non-lasting organic poisons extracted from the plant realm as an intermediate or conversion measure, it is important to know under what conditions the effects of these sprays are enhanced or counteracted. Nicotine, highly toxic in concentrated solution, should be mixed with soap or detergent solution in order to penetrate the waxy layer of the insect skin, especially aphids. Its residual toxicity is about one day to one week. It becomes more effective if hydrated lime is added, while rotenone and pyrethrum become ineffective in alkaline solutions.

Predatory insects are lady beetles and their larvae (against aphids and scale), lacewing flies and their larvae, (called aphid lions); syrphid fly larvae (against aphids); nocturnal ground beetles and their larvae; tiger beetles, the praying mantis, robber flies, predaceous bugs in general, and spiders. Shrews, skunks, bats, snakes, newts, lizards and toads are welcome in an orchard. Damaging insects, especially their larvae, can also be subject to parasites that eventually destroy them. Certain wasps lay their eggs into other larvae. The oriental fruit moth has a parasite, *Macrocentrus ancylivora*, and the egg parasite *Trichogramma minutum* attacks many undesirable insects. Orchardists do well to brush up on their knowledge of all these things. It is beyond the scope of this little book to present all the details.

Orchard Pest Management

Michael Maltas

Introduction

This orchard management schedule was developed while establishing a small commercial farm and homestead in the Missouri Ozarks, North America. This program has been developed primarily for use with apples in South Central Missouri. Adjustment for other locations and tree fruits should be made accordingly.

Four spray schedule treatment options are provided: Biodynamic, Organic A, Organic B, and Limited Synthetic. These options are consistent month to month and it is assumed that orchardists will choose the appropriate option at each successive stage. It is advisable for those choosing the limited synthetic option to adhere to that option throughout as the organic and biodynamic options rely on natural predator/parasite complexes which may be disrupted by synthetic applications.

It should be noted that, in this region, only a synthetic program, accompanied by regular scouting, weather monitoring and so forth, is likely to provide a consistently high percentage of commercially marketable fruit – 'organic' consumers being as cosmetic conscious as everyone else.

Even after several years of experimentation and adjustments only 20% to 50% marketable fruit is a reasonable expectation, in an average year. Also where single plantings exceed 50 to 100 trees, the non-synthetic options become less effective.

These comments are not meant to discourage growers from ecologically responsible practices but to impress upon them the challenges and difficulties of producing commercial tree fruits in this region.

Various sprays, pastes and preparations are mentioned throughout the schedule. Formulas and detailed explanations are in the chapter Formulas, Recipes, Chemicals (page 77).

No attempt has been made to justify or explain the specific biodynamic formulations and methods. This is too big a subject and only those familiar with the principles involved should really use these preparations. Caution is particularly advised with use of the 501 silica preparation. For those interested, a good starting point is your local Biodynamic Association (see Appendix for contact details).

There are a few references to moon constellations. During the course of a (sidereal) month the moon passes through the twelve constellations of the zodiac. The constellations are related to the classic elements of earth, water, air and fire. Details can be found in the annual *Maria Thun Biodynamic Calendar.*

Read this first!

To use this schedule effectively you are going to have to think! It is a very condensed version of my experiences and research. Much is left out, much is unfinished; in addition only those controls that do work, have worked in some situations, or have a potential to work, are listed here. There is loads of stuff that did not work and I do not have time to go into it all. All the usual references to rootstocks, cultivars, pruning methods, equipment, ag-chem suppliers, etc., have been left out. This material can be found in numerous other publications and online.

Most importantly, there are some specific recommendations referred to that I realise from initial feedback, have been 'too condensed'! In other words, to be effective, more data is necessary, more testing is needed, or they have possible injurious side-effects yet to be determined. In addition, growers in some regions may wonder at the surprising lack of information given to some 'major' pests. Likewise, those interpolating to other climates (particularly the dry West) need to be aware of unexpected differences.

If something does not work for you, find out why and you will automatically have deepened your understanding of the problem and the possible controls in the process!

Some points to watch for in particular are listed below.

Plum curculio

Because curculio is listed as a pest in this schedule, there may be the implied expectation that the organic controls listed will nail it. This is not true; or rather, not realistically true. The botanicals available to date have not shown adequate control when this pest decides to move in. Sure, super strength Rotenone-5 plus Pyrethrum and loads of 'synergists' may knock it down for a day or two, but what about a consistent period of six weeks? Consider the expense of such super-frequent sprayings in money, time, massive compaction to soil from rigs and equipment, annihilation of all beneficials, orchardist burn-out, etc.

To our surprise we did not seem to see any evidence of curculio over three cropping years in our orchard; maybe it was too young, maybe our 'isolation' factor helped, maybe they didn't like our biodynamic methods, maybe we were lucky?!

Also related to curculio: the 'green apple curculio traps' mentioned on pages 60–61 *only* work if there is an apple-scented perfume (made by Avon) used in conjunction with them. Vials are hung in the tree under or near the traps.*

* Leblanc, Jean-Pierre R., et al. 'Oviposition in Scout-Apples by Plum Curculio', *Environmental Entomology*, 13:286-291. 1984.

Tarnished plant bugs

These guys are also very difficult to control with botanicals (same reference as curculio above).

Codling moth

Our major surprise number two! We also had no evidence of codling moth; neither did a larger organic grower (12 acres of apples and pears) in the same bio-region. Hence, I don't give it much print.

For those who do have problems with it, Ryania is the classical control, as well as use of superior oils, key timed *Bacillus thuringiensis* (Bt), Trichogramma, mass-trapping, etc. In relation to the codling moth virus, about all you'll need to know is in the *IPM Practitioner.**

Borers

Borers can be a major pest and very difficult to deal with; wire, knife and probing their tunnels may become your only option. Certainly they tend to go for weakened or stressed trees first; however, your whole site may be 'stressed' by soil, climate, management, or other factors.

Newly cleared forest land and immediate tree planting thereafter could be an open invitation to trouble. Pulling the land out of a 'forest condition' with a year or two of other crops (or green manures), then planting the trees, may help a lot.**

* 'Update: Codling Moth – Big Changes Ahead', *Integrated Pest Management (IPM) Practitioner,* VII.5. May 1985.

** See also the excellent article on borers by John Hillbrand, *Pomona,* (journal of NAFEX, North American Fruit Explorers) Spring or Summer 1987.

Some warnings

A.P. Thompson's mixture (Borer spray 1, page 77) with moth crystals could be fatal to young trees! A.P. assured me that it worked for him (on older trees?) but I know of no one else who has used it. John Hillbrand reports killing back on young apple trees with use of moth crystals, used in the same fashion as that recommended since the 1930s for peach trees and peach tree borers.

Mint as a repellent has worked against peach tree borer, a moth. But I have no evidence that it is effective against the beetle group; for instance, round-headed and flat-headed apple tree borers, etc.

Regarding the creosote repellent I would suggest extreme caution on young trees, and get the natural creosote extract if you find it. Since creosote was banned as a suspected carcinogen (keep off your skin), this option may be hard to follow.

Tobacco dust needs to be applied *before* egg laying (emergence) and may only work against the moth group; for instance, peach tree borer, etc.

Soap repellents

Bars of soap are mentioned as repellents for deer and possibly borers. However, those in dry western climates in particular, need to be aware that it is the continual wetting of the soap that seems to make it effective. Once it has thoroughly dried out (or desiccates in very cold temperatures) it loses its protective value. It seems to work best in humid conditions above freezing point.

Pheromone traps

Watch out, you may invite more of the bad guys in than you had before!

Suggested Schedule

January – February

Dormant period

Expected problems

- Rabbits, especially when snow is on the ground
- Sunscald, especially on southwest side of tree trunks

Potential problems

- Deer
- Voles

In late February you can start pruning apples (and pears) during mild weather. To enhance vegetative growth, prune now and early March. To get minimal vegetative growth, dormant pruning should be done as late as possible – just before silver tip.

If sunscald or rabbit damage has occurred, use light colored tree paste to cover injury areas.

Spray options

Biodynamic

Spray the crown and trunks of trees with tree paste spray anytime between February and late March (after dormant pruning if appropriate). Spray on a dry day so the film on the trees can dry off well.

Organic A

No recommendation.

Organic B

No recommendation.

Limited synthetic

No recommendation.

March

Late dormancy – tree activity can begin in early years

Expected problems

- Voles
- Rabbits
- Sunscald

Potential problems

- Deer
- Aphids
- Mites
- Scale
- Scab
- Fire blight (preventive action period)

Main dormant pruning period. Put matured compost around crown drip-line of trees if not done already in fall (fall *is* the best time for this task).

Spray options

If silver tip and green tip commence in March, proceed immediately to the March-April spray recommendations.

Before using any copper based materials or sulfur products as outlined henceforth, see Formulas, Recipes, Chemicals for cautions and an understanding of their actions and limitations (in particular, reference to Bordeaux, Burgundy, dormant oil, dormant sprays, lime sulfur)

Biodynamic

Spray tree paste if not done in February, or repeat application on crown and trunk of trees after pruning is complete, if desired.

Spray Equisetum tea 508 on orchard floor late in the month (or earlier, if the season is advancing rapidly) preferably just before full moon.

Organic A

Several Choices:

1. Lime sulfur or Bonide oil and lime sulfur spray primarily for scab; has no strong bactericidal action like the copper sprays. Also destroys some beneficial mites and insects. This spray will not control fire blight.
2. Burgundy mixture is supposedly safer than lime sulfur for beneficials. Essentially fungicidal for scab, mildew cankers, etc.
3. Bordeaux mixture has both fungicidal and bactericidal properties. May cause russeting, especially if used later than dormant. Can be phytotoxic on some varieties in some seasons. Good for fire blight and scab, etc. Controls peach-leaf curl in peaches.
4. Dormant oil has a smothering effect on scales, mites and aphid eggs, etc. Can be combined with Bordeaux and sprayed at green tip. Do not apply at temperatures below 40° F (4° C).

Organic B

Same as Organic A.

Limited synthetic

Same as Organic A.

March–April

Bud break to bloom period

Expected problems

- Deer
- Voles
- Rabbits
- Fireblight
- Scab
- Powdery mildew
- Aphids

Potential problems

- Mites
- Scale
- Leafrollers
- Plant bugs
- Curculio

Finish pruning and 'detail' work on trees. Apply dormant spray option if not already done. As soon as the soil is sufficiently dry and tree activity begins, cultivate weeds.

Spray options

Biodynamic

As soon as trees and grass begin to grow, spray biodynamic preparations 500 and 507 Valerian on soil surface on a warm day in the late afternoon. This is preferably done shortly after weed cultivation.

At green tip to ½ in (1 cm) green spray 508 Equisetum tea plus seaweed foliar, and/or Scab spray of KOLO sulfur-clay preparation, wettable sulfur.

Between ½ in (1 cm) green to bloom possibly spray of 505 on orchard floor.

Equisetum and possible sulfur compounds are best applied as 2 separate sprays with several days interval between applications. Sulfur is stronger than equisetum and will override its action.

Organic A

Green tip to ½ in (1 cm) green: delayed dormant oil (1%) plus Bordeaux (6-6-100) (or you can substitute other copper compounds). This may give residual protection for up to 20 days for scab and fireblight, etc., if a slow season. However, if a lot of new tissue is emerging, these new leaves will not be covered by the protective spray (fungicides are preventative, not curative). So remain aware of development stages, speed of growth, etc., and be ready for another copper spray if necessary. *Always spray coppers when plants are in a dry condition* to reduce phytotoxicity risks.

Use *Bacillus thuringiensis* (Bt) for caterpillars. Do not tank mix with bactericidal compounds, etc.

Use Pyrethrum/Rotenone if caterpillars are associated with aphids.

Organic B

Green tip to ½ in (1 cm) green. Use lime sulfur with caution if dormant oil has been sprayed in the last three weeks, and temperatures remain below 85° F (30° C). Spray under dry conditions. Use the weaker strength, non-dormant rate.

Apply *Bacillus thuringiensis* (Bt) or Pyrethrum/Rotenone as indicated in Organic A.

Limited synthetic

Green tip to ½ in (1 cm) green. Use Polyram at label rate.

Streptomycin 17% for fireblight control.

Important: Do not use soap sprays at this stage when you want materials to stick to the leaves. Soap action may emulsify and lift off spray materials with the first rains.

Aphid control: if you have *not* used a smothering spray with dormant oil or particularly delayed dormant oil, an aphid spray *may* be needed. Scout, using a magnifying glass.

Do not use soap products such as Safer's if you've recently sprayed materials that may wash off (for instance, oils, etc.). If not, spray soap just before a rain so that the soap washes off quickly. A better option is to use Rotenone 5 or an Equisetum/ diatomaceous earth aphid spray, or 24-hour nettle brew or potassium permanganate spray.

April–May

Bloom, petal fall and 'early covers'
Pollination occurs during this period. Take extra precautions with all sprays, especially insecticides toxic to bees

Expected problems

- Fireblight
- Scab
- Powdery mildew
- Cedar apple rust
- Quince rust
- Aphids
- Plant bugs
- Curculio
- Leafrollers
- Codling moth

Potential problems

- Deer
- Voles
- Rabbits
- Mites (if warm enough)

Scout for insects, etc. Depending on season (slow or fast) *may* need additional fungicide cover, but is often not necessary.

Spray options, ½ in (1 cm) green to tight cluster

Biodynamic

Possible spray of 501 *if no danger of frost* (experimental only!).

Organic A

See March–April (not necessary unless missed previous spray).

Organic B

See March–April (not necessary unless missed previous spray).

Limited synthetic

See March–April (not necessary unless missed previous spray).

Spray options, tight cluster to first pink

This is the most critical period for powdery mildew control. Codling moth traps should be hung.

Biodynamic

Pre-bloom spray. Use either the 'dry weather' or 'wet weather' mix depending on rainfall and humidity. If frost danger exists, spray with 507 Valerian.

Organic A

If good cover has already been achieved with the Bordeaux and dormant oil, no additional protection may be necessary at this stage.

Organic B

Continue green tip cover spray (See March–April) every five to seven days till bloom.

Limited synthetic

Polyram or Thiram (at label rate). Streptomycin 17% for fireblight.

Spray options, pink

Tulips bloom just before pink; this can be used as phonological indicator. This is the critical rosy-apple-aphid period. If temperatures exceed 65° F (18° C) it is also critical for plant bugs and plum curculio.

Inspect codling moth traps regularly. IPM data exists regarding flight versus temperature at sundown.

Put up apple maggot and plum curculio traps: red sticky balls; yellow sticky cards; green store-bought apples and tangle foot for curculio.

Scab infections are likely if leaves are continuously wet for 15 hours at 50° F (10° C) or 9 hours at 62° F (17° C) or above.

For insect problems use *Bacillus thuringiensis* (Bt) or appropriate botanical for all options.

Biodynamic

Same as for tight cluster.

Organic A

Same as for tight cluster.

Organic B

Same as for tight cluster.

Limited synthetic

Same as for tight cluster. However, check label directions for residual period and determine days since last treatment. This application may not be necessary.

Spray options, bloom

Use no sprays except 'mild' materials which will not affect pollinators.

Mow dandelions, crimson clovers, etc., on orchard floor to curtail competitive blooms during this period.

Biodynamic

Spray 500 and 507 on the orchard floor immediately after mowing or during this period.

Spray 507 Valerian in crown of trees if frost threatens.

For fungicidal action spray Equisetum tea 508, seaweed foliar, Valerian 507.

Organic A

Aphids (in crown): potassium permanganate application in evening. *Do not* use soap spray.

Woolly apple aphids at soil line, if a problem, spray either Safer's soap or Nasturtium woolly apple aphid spray.

Caterpillars: use Bt.

For fireblight (experimental option only): use chlorox bleach solution.

Organic B

Same as Organic A.

Limited synthetic

Polyram or Thiram (as labeled); streptomycin (as labeled).

Spray options, petal fall (90% of blossoms off)

This is a critical period for plum curculio and many other insects. Codling moth may be active. Aphids may be a problem if rainy or hot. Mow and cultivate weeds from here on as necessary. Begin training of young branches

Biodynamic

Use petal-fall spray. You can start using 501 periodically, at least once or twice during early spring.

Organic A

There are two alternatives.

Use Petal-fall spray only; *Bacillus thuringiensis* (Bt) can also be applied if necessary, but may be best as a separate spray.

Or to obtain fireblight and better fungicidal action, use Petal-fall spray *but omit the sulfur* and make a second spray application of Bordeaux mix, 6-6-100 if dry and clear, or 4-4-100 if humid (potential problem with russeted fruit). Bt *may* be applied also if necessary as a third separate spray.

Organic B

There are two alternatives.

Use Petal-fall spray only; *Bacillus thuringiensis* (Bt) can also be applied if necessary, but may be best as a separate spray.

Or, to obtain better fungicidal action (but no fireblight control), use Petal-fall spray *but omit the sulfur* and make a secondary spray application of lime sulfur at rate of 3 gal/100 gal or less (potential problem with russeted fruit). Bt may be applied also if necessary, but as a separate spray.

Limited synthetic

Imidan (full strength), Polyram or Thiram (label rate), Nu-Film 17 as spreader-sticker.

Spray options, first cover (7–10 days after petal fall)

If weather is dry, wait till 10–15 days after petal fall.

Biodynamic

Petal-fall spray, but reduce sulfur to 4 lb/100 gal or 2 tbsp/gal (5g/l).

Organic A

Petal-fall spray, but *omit sulfur*. Separate spray of lime sulfur at 2 gal/100 gal.

Organic B

As Organic A.

Limited synthetic

Imidan, Polyram or Thiram, Nu-Film 17 spreader-sticker.

Spray options, second cover (14–20 days after petal fall)

Biodynamic

Petal-fall spray, but reduce sulfur by half. If clear, dry weather, advance to use recommendation for third cover.

Organic A

Not necessary unless fireblight in evidence.

Organic B

Not necessary unless scab in evidence.

Limited synthetic

Not necessary unless scab or curculio in evidence.

Spray options, third cover (21–30 days after petal fall)

Biodynamic

Third cover spray (see Formulas, Recipes, Chemicals).

Organic A

Same as first cover spray/Organic A.

Organic B

Same as first cover spray/Organic B.

Limited synthetic

Imidan, Ferbam or Thiram, Nu-Film 17, spreader-sticker.

May–June

Early cover period moving into summer schedule

Expected problems

- Fireblight
- Scab
- Cedar-apple rust
- Powdery mildew
- Quince rust
- Curculio
- Codling moth
- Leafroller
- Leafminers
- Plant bugs
- Mites

Potential problems

- Apple maggot
- Borers

Training and pruning continues. Scout regularly for pests.

In the event of a late spring, early cover sprays continue into this period. Otherwise, summer cover sprays commence once the fourth cover is completed.

Spray options

Biodynamic

Spray 500 if only used once in spring so far. Spray 501 at sunrise. Summer cover spray every 10–20 days depending on weather.

Organic A

Wet weather: Keep on using Bordeaux, lime sulfur or sulfur based options. Do not use sulfur if temperatures exceed 85–90° F (29–32° C). Use every 10–20 days.

In dry weather switch to periodic Bordeaux if fireblight is a problem and use the Summer cover spray every 10–20 days.

Insects: based on scouting, apply *Bacillus thuringiensis* (Bt), Safer's soap or appropriate botanical.

Organic B

Same as Organic A.

Limited synthetic

Follow State University Extension recommendations (or similar local agricultural guidelines outside the US) for localised problems that may arise. Otherwise, continue use of insecticide/fungicide combinations every 12–24 days, depending on weather.

All synthetic spraying should be terminated well before harvest. Check labeling for guidelines.

June–July

Expected problems

- Codling moth
- Borers
- Grasshoppers
- Eastern tent caterpillar
- Leafrollers
- Leafminers
- Scale
- Mites
- Scab
- Powdery mildew
- Bot rot

Time for more training, thinning, summer pruning. Mulch, supplemental irrigation.

Observe grasshopper hatch. To control:

- Use Nosema locustae
- Grasshoppers 'roost' at night in tall grass and weeds, so mow orchard floor and perimeters very close to the ground leaving 'strips' of unmowed grass one mower-width approximately every 20 ft (6 m). Wait 24 hours and return at night or very early in the morning and mow the 'strips' (plus 'roosting' grasshoppers) in high gear.
- Poultry/guinea fowl/duck free range on orchard floor. Introduce poultry as soon as possible after grasshoppers hatch.*

* An excellent reference article on this subject may be found in the NAFEX publication *Pomona*, Spring 1986.

Allow beneficials to control aphids, mites as much as possible. Tentative control of borers:

- Use Borer control sprays.
- Plant mint around tree base as possible repellant.
- Tobacco dust and/or Moth flakes around trunk base as possible repellant.
- Use tree paste on trunks to prevent egg laying.
- Paint trunks with creosote as repellant (use gloves, keep off skin).
- Paint trunks with a white latex paint to prevent egg laying.
- Hang lye soap on the trunk from planting on.

To control egg laying with synthetics, check State Extension publications.

Once terminal growth has stopped and June drop and fruit-thinning period is over, the period of maximum fruit bud differentiation is underway.

Spray options

Regarding Biodynamic, Organic A, Organic B, this is where you have to decide how much time you want to spend on a spray rig. The summer cover sprays that follow until harvest should be the appropriate fungicidal, insecticidal and foliar feeding sprays as determined by orchard observation and weather. Interpolate from Bloom Sequence Calendar (Appendix) and previous spray options for controls for specific problems. Avoid broad spectrum tactics as much as possible!

Biodynamic

Three applications of 501 spray about 9–10 days apart on days when the moon is in a Warmth/Fire constellation.* (Be aware that fruit bud development occurs underneath the leaf rosette apparent at this time. Leaves of this rosette may drop at this point with transition to a fruit bud.)

Summer cover spray every 2–3 weeks depending upon weather and necessity.

Organic A

Summer cover spray every 2–3 weeks depending upon weather and necessity.

Organic B

Summer cover spray every 2–3 weeks depending upon weather and necessity.

Limited synthetic

Same as May–June.

* See, *Maria Thun's Biodynamic Calendar.*

August

Harvest begins

Expected problems

- Secondary scab
- Sooty blotch
- Fly speck
- Codling moth
- Leafrollers
- Leafminers
- Mites
- Grasshoppers

Irrigation may be needed. Apply appropriate natural insecticides as needed (scouting). Keep grass mowed to reduce drought stress on trees, to keep grasshoppers down and facilitate picking up dropped fruit.

Summer pruning – only if grasshopper pressure is not expected to be severe – may have to wait till mid-September.

Molasses 'sun-shield' spray of 5% or so if drought stressed (experimental only).

Spray options

Biodynamic

Spray 501 in the very early morning, one week before harvest; exercise caution if severely drought stressed.

User Summer cover spray as necessary. If trees go dormant, it is perhaps best to leave them dormant by not applying stimulating foliar feeds.

Organic A

Use Summer cover spray as necessary. If trees go dormant, it is perhaps best to leave them dormant by not applying stimulating foliar feeds.

Organic B

Same as Organic A.

Limited synthetic

Follow State Extension recommendations.

September

Harvest, fall preparation, second growth

Expected problems

- Deer
- Leafrollers
- Caterpillars
- Sooty blotch
- Fly speck

Potential problems

- Voles
- Codling moth

Various caterpillar species may be present to hasten leaf drop and hardening off. Large numbers however, may result in damage to fruit buds set for following year. Use *Bacillus thuringiensis* (Bt).

Fall-summer pruning around mid-month if avoiding regrowth phenomena which can force fruit buds lower down on scaffold to break, if too much remaining vigor.

Spray options

Biodynamic

Use 500 if you want to stimulate regrowth. May be best to wait until later in October.

Organic A

No recommendation.

Organic B

No recommendation.

Limited synthetic

No recommendation.

October

Leaf fall begins

Expected problems

- Deer
- Voles

Spray options

Biodynamic

Use 500 in mid-late October if not sprayed already in September. You can add some stinging nettle brew to the last part of the stirring.

Spray Equisetum tea 508 on the orchard floor as leaf fall starts.

Organic A

Experimental at 90% leaf fall: to possibly control scab spores on ground and leaves and possibly fireblight for next year, spray Bordeaux 6-6-100 or possibly copper sulfate followed by lime. Spray to runoff so trees and leaves on ground get some spray material. If scale was a problem, spray dormant strength lime sulfur or possibly Safer's soap.

Organic B

Same as Organic A.

Limited synthetic

Follow State Extension recommendations.

November–December

Expected problems

- Deer
- Voles

Mow orchard and perimeters as short as possible to give an 'open' winter orchard.

Cultivate to break up vole tunnels if appropriate or use poison bait stations. Early in November is time for vole preparedness. *Do not* use poison bait stations if poultry have access to the area.

Apply matured compost around drip lines of trees on an Earth/Root (moon constellation) day if possible.

Tree paste can be applied to trunks, scaffolds.

Spray options

Biodynamic

Spray 508 Equisetum on orchard floor if not done already in October. Spray 500 on a warm, still cloudy afternoon, if possible after the composting. Can use tree spray in December on a warm dry day if possible.

Organic A

No recommendation.

Organic B

No recommendation.

Limited synthetic

Follow State Extension recommendations.

Mammal Control

Deer

Electric fence: minimum wire spacing around 12 in (30 cm) with bottom wire of grounded barbed wire about 6 in (15 cm) off of the ground, and first electrified wire 12 in (30 cm) above that. Under deep snow conditions additional 'ground' wires may be necessary alternating with 'hot' wires as deep snow can prevent good grounding. Ground wires should be connected to a copper rod driven 4 to 6 ft (120–180 cm) into the earth.

Non-electric fence: field netting to thick mesh and a 'sloping-out' angle to 6–7 ft (1.8–2 m).

Hinder deer and rabbit repellent (use label rate).

Dial or Ivory Soap hung in trees. This seems to be very effective during wet weather when soap is regularly moistened – not a reliable control during long dry spells!

Pie pans on strings.

Bloodmeal: uneconomic and not effective over time, with heavy deer pressure, or in cold weather.

Urine plus bloodmeal in juice jars: for mild deer pressure in warm weather does help.

Egg spray: a few eggs blended and mixed with 1 gallon (4 l) water, spray on a dry day. (Can add Nu-film 17 as a possible extender.)

Rabbits

Tree collars (the plastic spiral type) are very effective, but remove in growing season if possible as they harbour insects and fungus, etc. against the trunk.

Half-inch (1 cm) hardware cloth is effective, but a perimeter fence can be cheaper if you have a lot of trees! Make sure the hardware cloth is 12 in (30 cm) or more above winter snowline. Even galvanised wire tends to corrode rapidly in acid soils.

Perimeter fence of 2 in (5 cm) chicken netting 24 in (60 cm) high, with rebar posts simply threaded through the mesh every 3 yards (m) or so, may be cheaper than hardware cloth. This is most effective and very satisfactory, and easily removed in spring. Reusable indefinitely if stored well.

Hinder deer repellent (follow label rate).

Egg-spray repellent

Urine plus bloodmeal in juice jars: more effective in warm weather (odor development). Helps, but is not an effective repellent in colder weather or on its own.

Tree paste plus eggs and bloodmeal is an effective repellent/protectant. Renew if heavy rains wash off over some weeks.

Voles

It is important to distinguish between surface-feeding meadow voles and underground pine voles.

Remove all non-gravel mulches in early fall. Apply a gravel mulch, 2 in (5 cm) thick, around young trees. Mulch should extend out about 2 ft (60 cm).

Crushed rock ¼–½ in (5–12 mm) diameter may be best, though caution is advised as sharp edges may damage trunks.

When planting or replanting, build a gravel cone of approximately ½ in (12 mm) diameter gravel from the top surface of the roots up to the trunk at ground level.

Cut grass and clean up orchard in fall. Cut grass as short as possible for maximum exposure.

During a dry spell in late fall, cultivate around trees with Weed Badger or similar cultivator to break up runs near surface. This though will destroy a surface gravel mulch.

Half-inch (1 cm) hardware cloth tree collars – for meadow voles only.

Keep cats or poultry in the orchard.

Poison bait stations: bait placed in 18 in (50 cm) lengths of 2 in (50mm) PVC pipes; covered with straw/mulch etc. Cut-in-half car tires placed cut edge down are also good bait stations. Rozol bait is supposedly biodegradable. No poison bait stations in areas open to poultry.

Use Novole root stock (very vigorous, for large tree size requirements only).

Maintain a wide orchard perimeter and keep well mowed – especially near forest border.

Vole susceptibility is greatest for young trees so protective measures are designed primarily for the early years.

Gravel mulches

River gravel can be used in place of crushed rock; though not as 'sharp', it seems to be effective if placed correctly and thickly enough. Average pebble size should not be too large near the trunk. Coarser gravel can be used for the surface mulch further out from trunk. The gravel 'cone' below ground level at the trunk may help to reduce chances of collar rot.

Formulas, Recipes, Chemicals

Bacillus thuringiensis (Bt)

Bt is marketed as Dipel, Sok-Bt, Thuricide, Biological Worm Spray, and others. See label rates for dilution. Spray in early morning or evening when leaves are dry; or when coincides best with caterpillar's eating schedule.

Always use extender-spreader-sticker, for instance Nu-Film 17. Nu-Film plus Bt is viable for 7 to 10 days or more.

Do not mix with bactericidal sprays (like copper-based compounds, bleach and possibly sulfur).

Avoid mixing with fish emulsion. We suspect a release of chlorine destroys organisms.

Bleach

Fireblight spray (experimental): ten per cent solutions of standard household bleach (that is, 10% of the 5–6½ % sodium hypochlorite solution which is in the bottle as concentrate).

Spray three times during bloom period.

Biodynamic preparations

For those familiar with biodynamic principles make your own, or purchase from your local Biodynamic Association.

Borer spray 1 *from A.P. Thompson*

Mix 1 lb crystal moth flakes with 1 gallon spray oil (1 kg with 8 l). Transfer to paint can and mix with paint can mixer until thoroughly dissolved. Mix this with three times as much water (3 gal, 25 l) and spray onto base of tree trunk.

Borer spray 2 *from Hugh Williams*

Mix:

- 1 qt (1 l) tree paste spray
- 4 tbsp (60 ml) Diatomaceous earth
- 6 tbsp (90 ml) 5% Rotenone
- 1 tsp (5 ml) Dipel
- plus Nu-Film 17 spreader-sticker

Efficacy is doubtful under borer pressure in the Ozarks of Missouri!

Bordeaux mixture

Bordeaux is basically copper sulfate ($CuSO_4$) and lime in water. Can be bought premixed or be homemade. Homemade is much more effective and much cheaper.

Very effective protectant for apples, pears (pome fruits) against scab and fireblight. Peaches for peach-leaf curl. Soluble copper is very effective defoliant especially for peaches, etc. Used with caution during growing season but more safely in dormant periods. Can cause russeting of fruit in some varieties and seasons and some leaf chlorosis/damage.

Research evidence suggests some frost protection from copper sprays due to reduction of ice nucleating bacteria. Degree of protection is approximately 1–3° F (1–2° C).

Dissolve copper sulfate and lime separately, then bring together at the last moment in greatest mutual dilutions – keep well agitated. The example below is for a 6-6-100 Bordeaux mix.

Mix 6 lb of copper sulfate in 2 gal of water (3.5 kg in 10 l). Add to 48 gal (240 l) of water and agitate well. Mix 6 lb of hydrated lime in 2 gal of water (3.5 kg in 10 l). Add 48 gal (240 l) of water to the water tank, and pour in the lime mix. This gives a total of 100 gallons (or 500 l). *Do not* mix the concentrates together.

Substitutes

Other fixed coppers can be used with similar effects, for instance, cupric hydroxide, basic or tri-basic copper, etc. (follow label directions!)

You can substitute Top Cop with sulfur, or other commercial coppers (see also 'Copper compounds' below.)

Formulations recommended

- Dormant or into bloom period: 6-6-100
- Dormant or into bloom period severe disease situation: 8-8-100 (5 kg)
- Green-tip Bordeaux: 6-6-100 plus 1% dormant oil
- Post-bloom Bordeaux: 6-6-100 no oil or 4-4-100 (2.5 kg) no oil

Cautions

Over many years of excessive use copper can build up to toxic levels (for some plant species) in the soil. This may be more likely with biologically inactive soils.

Unless there is contrary evidence, do not spray on scab and fireblight resistant varieties or, only spray in early season.

Do not tank mix with other materials (except oil at delayed dormant.)

Only apply copper compounds during dry weather as phytotoxicity is enhanced if spray cannot dry on leaves, as soon as possible. The copper forms a shield cover on the surface. If it remains in solution (due to rain, dew, fog, etc.) it has penetrative ability and can cause toxicity problems. Spraying in the morning to allow quick dry-off is best.

Spray to cover the foliage and fruit with a thin film and not have trees drip heavily.

Some cultivars are more sensitive to coppers than others. Some examples are:

Bordeaux sensitive	*Not very sensitive*
Baldwin	Alexander
Ben Davis	Esopus Spitzenberg
Gravenstein	Fall Pippin
Jonathan	Hubbardston
R.I.G.	N-Spy
Twenty Ounce	Red Astrachan
Wagener	Red Canada
Wealthy	Rome
Yellow Newtown	Roxbury
Yellow Transparent	Tolman Sweet
	Tompkins King
	Yellow Bellflower

Botanicals

- Rotenone (1% and 5%), Pyrethrum, Ryania, Sabadilla, others.
- Rotenone/Pyrethrum mixes, for instance Bonide, Red Arrow.
- Rotenone/Pyrethrum/Ryania mixes, for instance, Tri-excel, Triple-plus.

Use label rates. Note that some materials may not tank-mix well with others.

Shop around, these insecticides are expensive and *often ineffective* – killing all the beneficials and leaving the 'bad guys' alive and well!*

Burgundy mixture

This is a copper-based spray widely used in Europe. It is possibly a safer substitute than lime sulfur for general fungicide use. And it should also have bacterial qualities.

Dissolve 3 oz of copper sulfate ($CuSO_4$) in a gallon of warm water (20 g/l). This may take several hours depending on type of copper sulfate. When fully dissolved, mix 4 oz of washing soda (sodium carbonate, Na_2CO_3) in a gallon of cold water (30 g/l), mix the two and spray *immediately*.

The above makes 2 gal. To make 100 gal, use 9½ lb copper sulfate in 50 gal, and 12½ lb washing soda in 50 gal (5.5 kg $CuSO_4$ in 250 l, and 7.5 kg Na_2CO_3 in 250 l, to make 500 l mixture.)

This has been used in dormant stage, emerging tissue and in full leaf stages of growth.

Well suited to pears, as most pear varieties are somewhat sensitive to sulfur and cannot tolerate lime sulfur in some locations and under humid conditions.

Copper compounds

Copper compounds have both fungicidal and bactericidal properties which make them well worth considering in any spray program. However, to avoid or reduce the possible danger (phytotoxicity, chlorosis, russet, fruit cracking, etc.) when used on green tissue an understanding of the action of copper may help.

* For additional information on botanicals and possible effectiveness on different insects, see 'Low Spray Plan For Apples' in *Organic Gardening*, May 1986, p. 88.

Phytotoxicity is apparently possible with *all* copper products under certain circumstances; however, some of the more 'modern' copper compounds like Top Cop with micronised sulfur, Kocide (cupric hydroxide), etc., claim to be 'safer' than Bordeaux. (Once in widespread use, Bordeaux has been replaced by the array of synthetic fungicides for scab, rust, etc., though is making a comeback as a bactericide as more and more antibiotic resistant fireblight bacteria show up.)

When on plant tissue, coppers must get 'fixed' out of solution as quickly as possible. Hence, spraying on damp days when drying would be slow is not recommended (for post-dormant, at least). Also, higher concentrations of copper tend to increase chances of damage, in general, but not always. However, making the solution too weak reduces effectiveness, so a balance is needed.

From this, a number of parameters are worthy of note. If the copper is very well fixed (as in tri-basic copper sulfate) then it has no or only limited 'penetrative' ability into the plant tissue (and so no damage is likely), but it may be too fixed to penetrate the fungus spores, cankers and bacteria. With Bordeaux, it's only after the (basic) lime in the solution dries off that the copper comes out of the soluble copper sulfate form adequately enough to be safe. In something like Top Cop ($CuSO_4$ with micro sulfur), it is the micronised sulfur that lowers the pH on the leaves that then (due to oxidation of sulfur to sulfuric acid) activates the fixed, tri-basic copper, and releases it sufficiently to kill bacteria and fungus spores. Hence in hot weather there is more sulfur oxidation and thereby more release of copper to do its work – but apparently not enough to cause excessive mobility of the copper with concomitant tissue damage to the plant.

From the above, one may speculate or conclude that in cooler weather or in non sensitive-tissue times, for instance, during dormant, early spring, or green tip time, a 'stable' form of copper will not be as effective as a more 'soluble' form, whereas in humid, muggy, late spring, we may be able to reduce the risk of copper by substituting the Bordeaux for something like Top Cop. Again in late fall if trying the late copper spray, a more 'soluble' copper seems to be a better choice – for who cares in November if any remaining leaves get 'bronzed' or phyto'd: the season is over. Using a less penetrative, 'stable' copper at this stage may only be 10% as effective on the overwintering 'bad guys' you are trying to get at.

In short, like most things, copper compounds need to be used intelligently!

Diatomaceous earth (kieselguhr)

Diatomaceous earth is the fossilised remains of microscopic marine animals. It works by scratching and wounding an insect's exoskeleton, causing dehydration.

Do not use synthetic diatomaceous earth as made for pool filters: it won't work as it is not abrasive.

Dormant oil

This is a broad spectrum suffocant. True dormant application should be 3 to 4 weeks before green tip; use tree paste spray instead if no mites or scale problems are anticipated. Do not apply if colder than 40° F (4° C), as it may not spread easily enough.

The label rate is 3 gal/100 gal, or 3¾ fl oz/gal (30 ml/l).

If not applying delaying dormant oil at green tip stage, apply at least one week before green tip, that is, around dormant to silver tip period.

Caution

There is a possible toxicity relationships between sulfurs and oils (see Sulfur compounds).

Dormant sprays

True dormant spraying is done around 3 to 4 weeks before green tip. Dormant sprays should be selected according to what needs controlling. If fireblight is a potential problem, Bordeaux or Burgundy should be used now or later at green tip. It is possible to use a true dormant spray of oil and Bordeaux later on.

Unless needed, avoid the use of lime sulfur and possibly even the dormant oil due to their potential impact on beneficials.

Use of the tree paste spray may well substitute for dormant oil if monitoring or prior experience gives no indication of serious pest problems (like mites, scale, aphids, pear psylla).

Equisetum/diatomaceous earth aphid spray

It is preferable not to use this if ladybugs are feeding on aphids as Rotenone/Pyrethrum will kill them.

To make 1 gal (4 l):

- 1 tsp (5 ml) Bonide Liquid Rotenone/Pyrethrum
- 2 tsp (20 g) Diatomaceous earth
- 1 pt (500 ml) Strong equisetum concoction

To make 100 gal (500 l):

- 1 pt (500 ml) Bonide Liquid Rotenone/Pyrethrum
- 4 lb (2.5 kg) Diatomaceous earth
- 12 gal (60 l) Strong equisetum concoction

Equisetum tea or decoction 508

If using dried *Equisetum arvense* (horsetail), cover with rainwater (unchlorinated) and allow to stand for several hours. Then bring to boil and simmer for half an hour to an hour. Allow to cool and strain through fine screen. This is the concentrate. Store in glass jars in a cool but light place; best not to keep over one month.

Tree paste

For use in tree paste dilute the concentrate with an equal amount of water, and use this to liquefy the paste to a proper consistency.

For 508 spray

Dilute the concentrate with 5 to 10 parts of rainwater, stir thoroughly for 10 minutes or so. Tepid water is better than very cold water.

Use as ground spray whenever necessary. Best apply in the afternoon around full moon.

Spray on leaves/trunk, etc., in dry (warm) conditions so that it can stick on the surfaces. Hence, before rain, etc., rather than during moist weather. Best applied when the moon is in Water constellations. Spray on leaves, trunk, crown mornings or afternoons.

Lime sulfur

See also 'Sulfur compounds' for more information. Used in different concentrations for winter dormant and summer growing seasons. Also homemade and commercial strength may differ; check label rates.

Do not use spreader-sticker.

Commercial grade rates for dormant application: 3–10 gal/100 gal (Missouri recommends 10%, that is, 10 gal/100 gal).

Commercial grade rates for summer (non-dormant) covers: less than 3 gal/100 gal (Missouri recommends 1%).

It is very effective against scab, fungus, etc., but also strong miticide and scalecide. May be used on green leaf tissue at lower concentrations early in the season if no post dormant oils have been used. Repeated use can cause leaf damage and sometimes fruit damage (cracking).

Cautions

- Will stain or bleach painted surfaces if unprotected from spray.
- Do not apply within 3 weeks of an oil spray after green tip.
- Do not mix with any other materials.
- Do not use when temperatures exceed 85° F (30° C).
- Do not use on sulfur sensitive varieties.
- Should not be sprayed after flowering on those trees not sprayed with it before flower (perhaps a tolerance condition).
- Use lower concentrations in humid climates.
- Do not use on peach foliage.

Nasturtium wooly apple aphid spray *from Hugh Williams*

If you cannot plant nasturtiums at the tree base, make a pressed extract of nasturtiums and dilute with water to make a 3–5% concentration of liquid. Spray at the tree base where the aphids congregate.

Safer's soap may also work. (Repeat applications if necessary).

Nosema locustae

This protozoan infects over 60 species of grasshoppers, mormon cricket and black field cricket.

Petal-fall spray

Mix following:

- Rotenone 5% (as labeled)
- Liquid Pyrethrum (as labeled)
- Ryania (as labeled)

- Diatomaceous earth (2 tbsp/gal, 4 lb/100 gal, 5 g/l)
- Spreader-sticker (as labeled)

Optional additions:
- Sulfur (4 tbsp/gal, 8 lb/100 gal, 10 g/l) dependent on options chosen. Reduce to half strength for later cover sprays
- Cryolite (as labeled), a natural mineral, effective on chewing insects. (Kryocide is a good synthetic equivalent)
- *Bacillus thuringiensis* (Bt) (as labeled) should not be necessary
- Quassia (as labeled)

Pheromone and other insect traps

For accurate monitoring of many insect pests, numerous traps like the red sticky spheres, yellow sticky cards, commercial or home-made pheromone traps, can be a valuable aid. In small orchards 'mass trapping' can even decimate localised pest insects (of some species) enough to be an effective control in itself. The more 'isolated' your site the more likely this method could succeed and usually two traps are needed per mature tree.

There are improved traps like biolure traps and a range of pheromone baits with a unique slow, continual and stable release rate of the attractant. This gives a longer lasting and more consistent release which should enable better and more accurate monitoring.

If going into trapping in a big way, it would be worth learning how to make your own traps and comparing different ones to see which is most cost effective under your conditions.

Potassium permanganate

This aphid and mildew spray is from Europe. It biodegrades quickly and is easy on beneficials. Use in the evening to spare bees.

Mix 1/4 oz/gal (1 1/2 lb/100 gal, 180 g/100 l)

Can be used at double concentration for control of damping off in seedlings as a soil fumigant.

Prebloom spray

Dry weather mix

- Tree paste spray (26 fl oz/gal, 20 gal/100 gal, 200 ml/l)
- Equisetum tea 508 (1 pt/gal, 12 gal/100 gal, 120 ml/l)
- Seaweed and prediluted fish emulsion (as labeled)
- *Bacillus thuringiensis* (Bt) if necessary (as labeled)
- Diatomaceous earth (4 tbsp/gal, 8 lb/100 gal, 10 g/l)
- Milk (1 cup/gal, 6 gal/100 gal, 60 ml/l)
- Valerian 507 (trace)

Wet weather mix

Lime sulfur, flowable sulfur or WP sulfur plus spreader-sticker.

In addition, applied as separate spray:

- Diatomaceous earth (4 tbsp/gal, 8 lb/100 gal, 10 g/l)
- Dipel (as labeled)
- Seaweed and prediluted fish emulsion (as labeled)
- Milk (1 cup/gal, 6 gal/100 gal, 60 ml/l)
- Valerian 507 (trace)

Safer's soap

This spray is primarily for aphids, but also used for mites, scale, etc. There is no residual action worth speaking of; it is a one-time effect per spray.

Best used later in season to avoid washing off the more critical early cover sprays. Do not spray during blossoming.

Use label rates (approx. 6 tbs/gal, 2½ gal/100 gal, 25 ml/l).

Seaweed

Seaweed's claims for disease resistance probably result from increased plant health. It contains growth regulators, particularly cytokinins, micronutrients, etc. Available in various brands including Maxicrop, Sea crop, Wachters.

Streptomycin

Depending on your perspective and that of your organic market – this product may well be accepted as a biological and therefore usable by organic growers.

Use label rate or Streptomycin 17% at ½ lb/100 gal (60 g/100 l).

Sulfur compounds

Sulfur and sulfur formulations have long been used against scab as an 'organic' fungicide though their action is not just limited to this role and there are many plant species (and even varieties within species) that are sensitive to sulfur and are damaged by it to various degrees (for instance, apricots, melons, squash, cucumbers). Familiarise yourself with these before applications.

Note that continued application of lime sulfur on green tissue has a good chance of causing leaf burn and sometimes cracking of fruit.

Sulfurs are also effective against powdery mildew and brown rot and have additional strong miticidal action plus other insecticidal qualities. Beneficial mites and insects can be negatively influenced by its use.

General principles

Lime sulfur (calcium polysulfide) is much stronger than plain sulfur (WP, dust, or flowable) in the immediate sense, but has far less 'residual' action. In other words, it is used primarily in the dormant or early bud break stages, or in weaker concentrations once green leaf tissue is in abundance for kickback action. It can reduce a scab infection that has already started, whereas plain sulfur alone is only prophylactic.

Plain sulfur (WP, flowable, etc.) needs to cover the leaves with a thin protective film before it infects scab. However, sulfur has much more sticking power (residual) than the lime sulfur. Hence they are often used in combination – the former for residual carry-over, the latter for immediate control.

Lime sulfur also seems to exhibit the puzzling phenomena of showing more possible phytotoxic damage if used post bloom if it was *not* used on the same tree *before* bloom!

Sulfurs and sulfur compounds should all be used with extreme caution at higher temperatures. Generally 85° F (30° C) is considered to be the cut-off point. Sulfur volatilises and 'fumes' at higher temperatures and this enhanced activity greatly increases the chance of phytotoxicity, leaf burn, russeting, etc.

In addition, humidity and light intensity play a role, as exhibited in the sulfur-oil cautions and differing label rates in the humid and dry areas (like eastern and western parts of the United States).

In general, oils, whether dormant or summer types, and sulfurs are a dangerous combination once green tissue appears and temperature increases. Lime sulfur and dormant oil can be used in combination even in the fully dormant stage or up to green tip, but after that they should not be used within 3 to 4 weeks of each other. Similarly, plain sulfurs should not be used with an oil or within 4 weeks of an oil application unless in dormant, delayed dormant, or post-harvest (all fruit off) applications.

Higher temperatures, stronger light, higher humidity and probably numerous other factors like drought stress, soil quality, nutrient uptake, demand caution. Isolate some test trees and find out what the sulfurs do in your location, under your conditions and concentrations before use on a larger, commercial scale.

As you can see, selection of 'disease resistant' cultivars can save you a lot of sleep worrying about all this kind of stuff. It's not just the synthetic chemicals that can be complex!

Sulfur sprays

See also 'Sulfur compounds' above for more information. Sulfur sprays are available as WP sulfur, Kolo Sulfur-Clay, liquid flowable sulfur, micronised flowable sulfur, etc.

See labels for recommended dilutions. With WP sulfur and others, use a spreader-sticker (often already added by manufacturers.)

WP sulfur is used at concentrations of 4–12 lb/100 gal (2–6 tbs/gal, 500–1500 g/100 l)

Flowable sulfur is used from pre-bloom to petal fall at concentrations of 3–7 cups/100 gal (190–440 ml/100 l). For cover sprays use 2–4 cups/100 gal (125–250 ml/100 l).

Make paste before tank mixing. Use the lower rates in humid areas. Higher rates can be applied in dry areas.

Cautions

- Many fruits are sensitive to sulfur (for instance, light skinned apples).
- If not used at label rates, it may be ineffective – so follow label.

- More damage occurs in humid climates.
- Do not use at temperatures above 85° F (30° C) as foliage burning may result.
- Do not mix or spray within a short period of applying oils after dormant; a 3 to 4 week interval may be necessary.
- Sulfur has insecticidal/miticidal properties and can affect the beneficials.
- Does not control cedar apple rust.

Summer cover spray

Mix following:

- Tree paste spray (2–3 cups/gal, 12–18 gal/100 gal, 120–180 ml/l)
- Equisetum tea 508 (1 pt/gal, 12 gal/100 gal, 120 ml/l)
- Nettle tea 508 (1 pt/gal, 12 gal/100 gal, 120 ml/l)
- Milk (1 cup/gal, 6 gal/100 gal, 60 ml/l)
- Seaweed and prediluted fish emulsion (as labeled)
- Molasses (2½ tbsp/gal, 1 gal/100 gal, 10 ml/l) up to 10% experimental

Tangle foot or tangle trap

Tangle foot is sticky material used to trap insects. Tangle trap is more viscous and is used to paint onto traps.

Do not apply directly to tree bark – it will damage or girdle the tree. Young trees can be killed by it!

Third cover spray

Mix following:

- Tree paste spray (1 qt/gal, 25 gal/100 gal, 250 ml/l)
- Equisetum tea 508 (2–3 cups/gal, 12–18 gal/100 gal, 120–180 ml/l)
- Diatomaceous earth (4 tbsp/gal, 8 lb/100 gal, 10 g/l)
- *Bacillus thuringiensis* (Bt) (as labeled)
- Milk (1 cup/gal, 6 gal/100 gal, 60 ml/l)
- Seaweed and prediluted fish emulsion (as labeled)
- Spreader-sticker (optional)

Tobacco dust

General use in controlling soil insects. Use at base of tree, possible borer control. (At least one Arkansas grower has reported failure of this method.)

Tree paste

The paste is for hand application, as it is thicker than tree paste spray. For sunscald protection add more Calphos or other 'light' powder ingredient (light clay) to act as a light reflector.

Only apply when there is ample time to dry properly or rain will wash it off.

Basic recipe

- 2–3 parts bentonite clay
- 1 part fresh cow manure

The clay should be 'soaked' overnight first to absorb water fully. Drain excess. Cow manure is best fresh from pasture pies: can be screened if necessary.

50 lb bags of bentonite or similar clay are often available very cheaply at farm supply stores for use as pond sealing material.

Additions to basic recipe

Dilute all to proper consistency with Equisetum tea

- less than 20% fine screened sand
- less than 10% fine screened compost
- few % Calphos. Add dry powder to make mix more firm and light in color: good for sunscald protection acting as light/heat reflector.
- trace of seaweed powder
- trace of eggs
- few % diatomaceous earth
- trace of whey
- trace of bloodmeal
- biodynamic preparation 500
- biodynamic preparation 507
- trace of nettle tea
- less than 10% cornstarch or potato starch to thicken and add more sticking qualities, if necessary.
- Quassia, Rotenone, Pyrethrum, etc.

Tree paste spray

This is a dilute and very fine screened version of the basic tree paste, with additions being those most suited to spraying. The dilution is primarily with 508 Equisetum tea.

It may be regarded as a biological substitute for dormant oil, though unlikely to be as effective in problem conditions of scale, mites, etc.

The tree paste spray is for use in sprayers, but avoid those with roller pumps, as it will wear them out. A diaphragm pump is best. For simple screening of the mix, pour through a double layer of nylon stockings stretched over the tank fill-hole, dangling full length into spray tank.

Possible additions as necessary:

- Diatomaceous earth, for scale (4 fl oz/gal, 30 ml/l)
- Seaweed foliar material (as labeled)
- Eggs, a repellent factor (less than 1 per gallon, 1 per 5 l)
- Biodynamic preparation 500 or Maria Thun's Barrel Preparation (trace)
- Biodynamic preparation 507 (trace)

Appendix

Bloom sequence calendar

Observations of two years on Pod Farm
in South Central Missouri, United States

1984

Silver tip & green tip, some ½ in (1 cm) green	April 15	
		9 days
Full pink	April 24	
		6 days
Full bloom	April 30	
		7 days
Petal fall	May 6	
		Total 22 days

1985

Silver tip & green tip	March 20	
		9 days
Green tip, ½ in (1 cm) green & tight cluster	March 29	
		10 days
Tight cluster & pink	April 8	
		5 days
First bloom	April 13	
		7 days
Full bloom	April 19	
		5 days
Petal fall	April 24	
		Total 36 days

Weight and liquid conversion

Per US gallon	*Per 100 US gallon*	*Metric*
1 ½ tsp or ½ tbs	1 lb/100 gals	120 g/100 l
3 tsp or 1 tbs	2 lb/100	240 g/100
1½ tbs or ½ oz	3 lb/100	360 g/100
2 tbs or ⅔ oz	4 lb/100	480 g/100
2½ tbs or ¾ oz	5 lb/100	600 g/100
3 tbs or 1 oz	6 lb/100	720 g/100
4 tbs or 1⅓oz	8 lb/100	1.0 kg/100
5 tbs or 1⅔ oz	10 lb/100	1.2 kg/100
6 tbs or 2 oz	12 lb/100	1.4 kg/100
2½ tbs or 1¼ fl oz	1 gal/100	10 ml/l
5 tbs or 2½ fl oz	2 gal/100	20 ml/l
7½ tbs or 3¾ fl oz	3 gal/100	30 ml/l
10 tbs or 5 fl oz	4 gal/100	40 ml/l
6½ fl oz or ¾ cup	5 gal/100	50 ml/l
7⅔ fl ozs or 1 cup	6 gal/100	60 ml/l
10 fl oz or 1¼ cups	8 gal/100	80 ml/l
13 fl ozs or 1½ cups	10 gal/100	100 ml/l
15 fl oz or 2 cups	12 gal/100	120 ml/l

US area	*Metric area*
1 unit/acre	0.4 units/ha
2.5 units/acre	1 unit/ha
1 lb/acre	1.12 kg/ha
0.9 lb/acre	1 kg/ha
1 ton/acre	0.37 tonne/ha
2.7 tons/acre	1 tonne/ha
1 gal/acre	9.35 l/ha
0.11 gal/acre	1 l/ha

Bibliography

Cloos, Walther, *The Living Earth,* Lanthorn Press.

Colquhoun, Margaret and Axel Ewald, *New Eyes for Plants,* Hawthorn Press.

Conford, Philip, *The Origins of the Organic Movement,* Floris Books.

—, *The Development of the Organic Network,* Floris Books.

Klett, Manfred, *Principles of Biodynamic Spray and Compost Preparations,* Floris Books.

Koepf, H.H. *The Biodynamic Farm,* Anthroposophic Press, USA.

—, *Koepf's Practical Biodynamics: Soil, Compost, Sprays and Food Quality,* Floris Books.

Kranich, Ernst Michael, *Planetary Influences upon Plants,* Biodynamic Literature, USA.

Masson, Pierre, *A Biodynamic Manual,* Floris Books.

Osthaus, Karl-Ernst, *The Biodynamic Farm,* Floris Books.

Pfeiffer, Ehrenfried, *The Earth's Face,* Biodynamic Farming & Gardening Ass. USA.

—, *Pfeiffer's Introduction to Biodynamics,* Floris Books.

—, *Soil Fertility, Renewal and Preservation,* Lanthorn Press.

—, *Weeds and What They Tell Us,* Floris Books.

Philbrick, Helen and R.B. Gregg, *Companion Plants and How to Use Them,* Biodynamic Farming & Gardening Ass. USA.

—, *Gardening for Health and Nutrition,* Steinerbooks, USA.

Sattler, F. & E. von Wistinghausen, *Biodynamic Farming Practice,* Biodynamic Agric. Ass. UK.

Schilthuis, Willy, *Biodynamic Agriculture,* Floris Books.
Soper, John, *Biodynamic Gardening,* Biodynamic Agricultural Ass. UK.
Steiner, Rudolf, *Agriculture (A Course of Eight Lectures),* Biodynamic Literature, USA.
—, *Agriculture: An Introductory Reader,* Steiner Press, UK.
—, *What is Biodynamics? A Way to Heal and Revitalize the Earth,* SteinerBooks, USA.
Storl, Wolf, *Culture and Horticulture,* Biodynamic Farming & Gardening Ass. USA.
Thun, Maria, *Gardening for Life,* Hawthorn Press.
—, *The Biodynamic Year,* Temple Lodge Publishing.
—, *The Maria Thun Biodynamic Calendar,* Floris Books.
von Keyserlink, Adelbert Count, *The Birth of a New Agriculture,* Temple Lodge Publishing.
—, *Developing Biodynamic Agriculture,* Temple Lodge Publishing.
Waldin, Monty, *Monty Waldin's Best Biodynamic Wines,* Floris Books.
Weiler, Michael, *Bees and Honey, from Flower to Jar,* Floris Books.
Wright, Hilary, *Biodynamic Gardening for Health and Taste,* Floris Books.

Biodynamic Associations

Demeter International
www.demeter.net

Australia:
Bio-Dynamic Research Institute
www.demeter.org.au
Biodynamic Agriculture Australia
www.biodynamics.net.au

Canada: Society for Biodynamic Farming & Gardening
in Ontario
www.biodynamics.on.ca

India: Biodynamic Association of India
www.biodynamics.in

New Zealand: Biodynamic Farming & Gardening Assoc.
www.biodynamic.org.nz

South Africa: Biodynamic Agricultural Association
of Southern Africa
www.bdaasa.org.za

UK: Biodynamic Association
www.biodynamic.org.uk

USA: Biodynamic Association
www.biodynamics.com

Index

You may also be interested in...

Pfeiffer's Introduction to Biodynamics

Ehrenfried E. Pfeiffer

'A classic text by one of the earliest biodynamic farmers in North America … A very useful introduction.'

– SCIENTIFIC & MEDICAL NETWORK REVIEW

Ehrenfried Pfeiffer was a pioneer of biodynamics in North America. This short but comprehensive book is a collection of three key articles introducing the concepts, principles and practice of the biodynamic method, as well as an overview of its early history.

The book also includes a short biography of Ehrenfried Pfeiffer by Herbert H. Koepf.

florisbooks.co.uk

Weeds and What They Tell Us

Ehrenfried E. Pfeiffer

This wonderful little book covers everything you need to know about the types of plants known as weeds. Ehrenfried Pfeiffer discusses the different varieties of weeds, how they grow and what they can tell us about soil health. The process of combatting weeds is discussed in principle as well as in practice, so that it can be applied to any situation.

The Maria Thun Biodynamic Calendar

This useful guide shows the optimum days for sowing, pruning and harvesting various plants and crops, as well as working with bees. It includes Thun's unique insights, which go above and beyond the standard information presented in some other lunar calendars. It is presented in colour with clear symbols and explanations.

The calendar includes a pullout wallchart that can be pinned up in a barn, shed or greenhouse as a handy quick reference.

Floris
Books